Lessons from the War in Ukraine for Space

Challenges and Opportunities for Future Conflicts

ANDREW RADIN, KHRYSTYNA HOLYNSKA, CHEYENNE TRETTER,
THOMAS VAN BIBBER

Prepared for the Department of the Air Force
Approved for public release; distribution is unlimited.

RAND PROJECT AIR FORCE

For more information on this publication, visit **www.rand.org/t/RRA2950-1**.

Published by the RAND Corporation, Santa Monica, Calif.

Library of Congress Cataloging-in-Publication Data is available for this publication.

ISBN: 978-1-9774-1521-9

Cover: Support Forces of the Armed Forces of Ukraine photo.

About This Report

Space services and the disruption of these services have played an important role in the war in Ukraine. In this report, we offer an open-source analysis of events in the space domain and their associated takeaways as part of a larger project of identifying lessons learned from the war related to space. Activities conducted by Russia, Ukraine, and Western nations supporting Ukraine have emphasized the value of space services, including satellite communications; positioning, navigation, and timing; and intelligence, surveillance, and reconnaissance. The events of the war have also exposed vulnerabilities, such as Russian jamming of global navigation satellite systems and hacking of commercial satellite providers. These dynamics have shaped how the conflict has been fought and suggest insights into the potential role of space in future conflicts.

Funding

The research reported here was commissioned by U.S. Space Command and U.S. Space Force Space Operations Command, and jointly conducted within the Strategy and Doctrine Program of RAND Project AIR FORCE and within the International Security and Defense Program of the RAND National Security Research Division as part of a fiscal year 2024 project, "Learning Lessons for Space from the Ukraine War."

RAND Project AIR FORCE

RAND Project AIR FORCE (PAF), a division of RAND, is the Department of the Air Force's (DAF's) federally funded research and development center for studies and analyses, supporting both the United States Air Force and the United States Space Force. PAF provides the DAF with independent analyses of policy alternatives affecting the development, employment, combat readiness, and support of current and future air, space, and cyber forces. Research is conducted in four programs: Strategy and Doctrine; Force Modernization and Employment; Resource Management; and Workforce, Development, and Health. The research reported here was prepared under contract FA7014-22-D-0001.

Additional information about PAF is available on our website:
www.rand.org/paf/

This report documents work originally shared with the DAF on September 10, 2024. The draft report, dated February 2025, was reviewed by formal peer reviewers and DAF subject-matter experts.

The views expressed in this report are those of the authors and do not reflect the official policy or position of the Department of Defense or the U.S. Government. Review of this material does not imply Department of Defense endorsement of factual accuracy or opinion.

RAND National Security Research Division

NSRD conducts research and analysis for the Office of the Secretary of Defense, the U.S. Intelligence Community, the U.S. State Department, allied foreign governments, and foundations.

For more information on the RAND International Security and Defense Program, see www.rand.org/nsrd/isdp or contact the director (contact information is provided on the webpage).

Acknowledgments

We thank Gen Stephen Whiting for supporting the development of the project, as well as Maj Gen Samuel Keener, Maj Gen Devin Pepper, and Brig Gen Chandler Atwood for their feedback and sponsorship of the larger project. We also thank several interviewees who spoke to us for the research effort. Ann Dailey and Jordan Logue also provided inputs to this research effort. We thank Bruce McClintock for his comments and support of the project as the co-principal investigator. We especially appreciate thoughtful reviews from Colin Smith and Stephen Oliver.

Summary

Issue

Space-based services—and the denial of these services—have played an unprecedented role in the ongoing war in Ukraine. The role of space in the war offers lessons for how space might shape potential future conflicts and how the United States must prepare for events in the space domain in the future. In this report, we provide an open-source account of space activities throughout the war and extract relevant lessons for the national security community.

Approach

This report is organized along three mission areas that proved most significant in shaping the war in Ukraine: satellite communications (SATCOM); positioning, navigation, and timing (PNT); and intelligence, surveillance, and reconnaissance (ISR), including overhead imagery or radar. For each mission area, we identify the prewar capabilities within Ukraine and Russia, specify how these capabilities were employed or disrupted in the war, and determine any associated challenges or issues for the key stakeholders in the conflict. We rely on publicly available information, including published statements by Western, Ukrainian, and Russian officials; literature from Russian and Ukrainian defense enterprises; and open-source reporting.

Findings

The analysis of the use of space in the war in Ukraine yielded the following findings:

- **Key space services will likely be disrupted in future conflicts.** The war demonstrated the growing importance of securing and/or denying space services for modern warfighting. Cyberattacks, Global Positioning System (GPS) jamming, and other threats have significantly shaped the war, and the growing sophistication of both Russian and Chinese counterspace capabilities only increases the likelihood that the United States and its allies might face similar disruptions in a future conflict.
- **The widespread availability of space services increases the transparency of the battlefield.** Commercial imagery and declassified intelligence helped to provide a surprisingly timely warning of Russia's initial attack and improved situational awareness throughout the conflict. The widespread availability of commercial imagery or released national imagery will likely make future large-scale conflicts relatively more visible in open sources and to wider audiences.
- **Commercial space services bring value and vulnerability.** The war demonstrated the value and vulnerability of commercial space services. Ukraine's ability to leverage Western commercial services enhanced its warfighting and reduced the need to field its own

capabilities. Concerns over the reliability of Starlink emerged partly because of the unique, informal provisions of its services to Ukraine; space services are more typically governed by contracts. However, Taiwan and other countries' hesitations about SpaceX highlights how countries are likely to be cautious about depending on a single provider.

- **Future warfare will present unique challenges and opportunities across different space mission areas.**
  - **SATCOM:** The war in Ukraine has demonstrated the potential utility of proliferated SATCOM architectures for enabling terrestrial forces. The Viasat hack also revealed the vulnerability of any single provider.
  - **PNT:** The war in Ukraine has shown how space-provided global navigation satellite systems will very likely face continuing challenges from terrestrial jamming.
  - **ISR:** Commercial ISR products or those shared by other partners can substitute for national capabilities, but an overdependence on shared capabilities can reduce U.S. partners' freedom of action. The increasing capability and availability of commercial imagery could also benefit adversaries. Space capabilities, including ISR, will increasingly require careful attention to limit the risk of misuse or cyber penetration.

Recommendations

We make the following general recommendations:

- **Plan to leverage commercial space assets to support allies and partners leading up to and during future conflicts.** The war in Ukraine has demonstrated how leveraging commercial resources can benefit U.S. allies and partners engaged in conflict. The U.S. government and commercial space companies should build on this experience with wargames and other planning exercises to explore what and when commercial space services might benefit different partners.
- **Continue developing robust contract arrangements with commercial space providers and better articulate U.S. responses to potential interference.** The United States and its allies should work with commercial providers to anticipate how contingencies can affect future service provisions and to articulate clear commitments to respond whenever possible. These efforts would help assuage the mutual concerns of the United States, its allies, and commercial space companies.

We make the following recommendations for specific mission areas:

- **SATCOM:** The U.S. Department of Defense (DoD) should continue investing in proliferated and diverse SATCOM architectures. A variety of available communication pathways—including both military-owned systems and commercial services—provides critical resilience against potential adversary counterspace threats.
- **PNT:** DoD should continue exploring innovative PNT technologies—including alternatives to GPS—to ensure platforms and munitions remain effective against adversary electronic warfare interference.

- **ISR:** The United States should set expectations for its allies and partners about what kinds of ISR products—both commercial and government—will be available or shared in conflict. Efforts should also be made to develop strategies for preventing adversaries from obtaining access to commercial space services.

Contents

Lessons from the War in Ukraine for Space

Introduction

Space-based services—and the denial of these services—have played an unprecedented role in the ongoing war in Ukraine. Information collected from space, including from commercial providers, showed Russian convoys amassing at Ukraine's border and later moving into Ukraine. A Russian cyberattack against the commercial provider Viasat disrupted key communication services being provided to the Ukrainian government.[1] Starlink services, offered initially on an essentially philanthropic basis from SpaceX, became a critical enabler for the Ukrainian military and provided tactical communication for ground forces. Although the involvement of SpaceX and other commercial providers offered new capabilities, it also raised new issues around how these providers should respond if targeted and how to ensure their reliability in providing services as the war intensifies.

The role of space in the war in Ukraine presents important lessons for how space might shape potential future conflicts and how the United States should prepare for developments in the space domain. This report is part of a larger project to provide insight to the U.S. Space Force (USSF) Space Operations Command and U.S. Space Command on lessons related to space from the war in Ukraine and for potential future conflicts. In this report, we present an open-source account of space activities throughout the war and extract relevant lessons for the national security community. One particular theme from our analysis is the importance of demonstrating both the significance and vulnerability of space services provided by commercial entities. There are also important lessons for the different space mission areas, including the continual challenge of the adversary denying space-based communications and navigation services.

We begin this report with a brief discussion of our research approach, followed by a historical overview of both Russian and Ukrainian space capabilities and strategies since the end of the Cold War. In the subsequent sections, we then document the role of space capabilities and activities in the war in Ukraine, categorized by the mission areas deemed most consequential to the conflict: satellite communications (SATCOM); positioning, navigation, and timing (PNT); and intelligence, surveillance, and reconnaissance (ISR). In the final section, we conclude with findings and recommendations.

---

[1] Marisa Torrieri, "How Satellite Imagery Magnified Ukraine to the World," *Via Satellite*, October 24, 2022.

Approach

In this report, we present an open-source account of space activities—and their implications—in the Russia-Ukraine war that began in February 2022 with Russia's full-scale invasion of Ukraine. After providing some historical context about Ukrainian and Russian space capabilities, we organize our analysis along three mission areas—loosely derived from U.S. doctrine—that have proven most significant in shaping the war to date: SATCOM, PNT, and ISR, including overhead imagery or radar.[2] For each mission area, we identify the prewar capabilities within Ukraine and Russia, specify how the capabilities were reported to be employed or disrupted in the war, and discuss any associated challenges or issues that emerged for Russia, Ukraine, the United States, or other stakeholders in the conflict. In many cases, space is important for enabling other military capabilities, such as command and control or long-range strikes. Therefore, we include relevant assessments about such enabled capabilities within the respective sections of the given space mission areas.

We relied on publicly available information for our sources, including published statements by Western, Ukrainian, and Russian leaders and officials; literature from Russian and Ukrainian defense enterprises; and open-source reporting from major media outlets. Although we recognize that these sources might not deliver a full account of activities in the space domain, they are useful for capturing as much information as possible. In addition, we conducted several not-for-attribution interviews with Ukrainian and U.S. experts for context. Our research was largely completed in July 2024, and, although we included some updates during the review process, the report might not reflect subsequent events.

Historical Context

The role of space in the war in Ukraine stems from the prewar space capabilities of Russia and Ukraine. However, as we will discuss in this report, the war has also forced Ukraine to rely on space services provided by the United States, other Western partners, and private companies. Although Ukraine benefited from these external services, its own prewar space legacy likely shaped its ability to leverage them.

During the U.S.-Soviet Space Race of the Cold War era, the Soviet Union—which included both Russia and Ukraine—was a pioneer in spaceflight and exploration. The Soviet space program launched the first earth-orbiting satellite, Sputnik 1, and sent the first animal into orbit aboard Sputnik 2.[3] Russia inherited much of the Soviet Union's space industry and capabilities, and the legacy of the Soviet space program remains important today. President of Russia Vladimir Putin has repeatedly referenced Soviet pilot Yuri Gagarin's historic 1961 flight aboard Vostok 1, such as in

[2] Joint Publication 3-14 outlines different space mission areas. We focused on the three that were most relevant to open-source publications on space services in Ukraine. See Joint Publication 3-14, *Joint Space Operations*, Joint Chiefs of Staff, August 23, 2023.

[3] Elizabeth Howell, "Sputnik: The Space Race's Opening Shot," Space.com, September 29, 2020; Mike Wall and Meghan Bartels, "Laika the Space Dog: First Living Creature in Orbit," Space.com, January 14, 2022.

2021, when he stated that Russia "will always be proud that it was our country who paved the road to outer space."[4]

Ukraine played an important role in Soviet space programs, serving both as a research hub and as a host to such key enterprises as the KB Pivdenne or Pivdenne Design Office (also referred to as the Yuzhnoye Design Office in Soviet and Russian sources) and the State Enterprise *Production Association Southern Machine-Building Plant named after O.M. Makarov* or Pivdenmash (also referred to as *Yuzhmash*). These facilities produced satellites, rockets, and Soviet intercontinental ballistic missiles.[5]

Since the collapse of the Soviet Union, both Russian and Ukrainian space enterprises have been in steady decline. In Russia, government funding shortages throughout the 1990s severely limited investment in the space sector. The lack of funding was exacerbated by other issues, such as brain drain, corruption, and organizational obstacles, which continue to hamper developments in the Russian space enterprise.[6] Test flights of Russia's Angara rocket have been repeatedly delayed because of supply and technical issues, and the construction of Russia's Vostochny Cosmodrome has been complicated by numerous embezzlement scandals.[7] The failure of Russia's robotic lunar lander, Luna-25, in 2023 is emblematic of the depth of the challenges plaguing the post-Soviet Russian space industry.[8]

Ukraine's integration into the Soviet space program laid a foundation for further independent development, but Ukraine too has struggled to maintain its space industry after the collapse of the Soviet Union. KB Pivdenne and Pivdenmash continued to manufacture carrier rockets, satellites, and several other space-related products—including for the United States and for European companies—but on a significantly smaller scale.[9] On January 13, 2022—a month prior to Russia's full-scale invasion—a SpaceX Falcon 9 launched Sich-2-30, Ukraine's first new satellite in a decade.[10] The declared main mission of the satellite was to conduct environmental monitoring, although Ukrainian media also mentioned that it could be used for such military purposes as obtaining data on the

---

4 "On Gagarin Flight Anniversary, Putin Vows Russia Will Remain Space Power," Reuters, April 12, 2021.

5 Peter Finn, "Ukraine Trying to Take Off into Global Space Industry," *Washington Post*, November 20, 1998.

6 Bruce McClintock, "The Russian Space Sector: Adaptation, Retrenchment, and Stagnation," *Space and Defense*, Vol. 10, No. 1, Spring 2017, p. 6.

7 Rahul Rao, "Russia Launches Heavy-Lift Angara Rocket on 3rd Test Flight, but Misses Intended Orbit: Reports," Space.com, December 30, 2021; Matthew Bodner, "The Long Road to Vostochny: Inside Russia's Newest Launch Facility," *Space News*, January 30, 2019.

8 "A Failed Lunar Mission Dents Russian Pride and Reflects Deeper Problems with Moscow's Space Industry," Associated Press, August 22, 2023.

9 Hlib Husyev, "Following the USSR Trajectory—Ukraine: A Great History of Ukrainian Space, from Independence to the Present Day. People, Factories, and Launch Vehicles" ["За траєкторією СРСР—Україна: велика історія українського космосу, від Незалежності до наших днів. Люди, заводи та ракети-носії"], Babel, September 9, 2019; Pavlo Krasnomovets, "Ukrainian Space Enterprises Have Created Thousands of Rockets and Hundreds of Satellites. Why They Did Not Adapt to a World Without the Cold War" ["Космічні підприємства України створили тисячі ракет та сотні супутників для СРСР. Чому їм виявилося так важко пристосуватися до світу без холодної війни?"], *Forbes*, December 27, 2021.

10 Observing Systems Capability Analysis and Review Tool, "Satellite: SICH 2-30," webpage, last updated May 16, 2022.

movement of ships and equipment.[11] However, it remains unclear whether the satellite provides any useful data for military purposes.[12]

Ukraine also has pursued integration into the European civilian space program. In the early 2010s, Ukraine sought opportunities for scientific collaboration with European researchers, including through the Twinning and Copernicus programs.[13] In December 2019, the Ukrainian Regional Center of the European Union's Copernicus earth observation program was launched, facilitating access to data from Copernicus Sentinel satellites for both the public and private sectors in Ukraine.[14] In April 2022, Ukraine submitted an application to join the European Space Agency.[15] Despite these efforts, Ukraine's domestic space capabilities proved too limited for the critical role space would play in the conflict. Instead, Ukraine had to rely on communication, navigation, and imagery services provided by the U.S. government, other Western partners, and private space companies.

Satellite Communications

SATCOM has proven to be a particularly impactful space service for military operations since Russia's full-scale invasion of Ukraine.[16] Space-based systems—and the associated ground station infrastructure and end-user receivers—enable national and strategic leadership to maintain situational awareness and to convey their intent to operational commanders. As one USSF report explains, "At the tactical level, SATCOM provides critical, beyond line-of-sight connectivity for mobile forces, enables control of remote sensors and remote or in-flight weapons, transmits real-time battlefield intelligence, and ties sensors to shooters."[17]

Russian and Ukrainian SATCOM suffered from underinvestment following the collapse of the Soviet Union. However, after Russia's initial invasion of Donetsk and Luhansk in Ukraine in 2014, Ukraine began leveraging commercial capabilities to enable its operations. In an attempt to disrupt these services, Russia sought to deny SATCOM by hacking the commercial provider Viasat. Since the 2022 full-scale invasion, Ukraine has also been enabled—and, to a degree, restricted—by SpaceX's provision of Starlink services.

---

[11] "Ukrainian Sich-2-30 Satellite Successfully Launched into Orbit," Defense Express, January 14, 2022.

[12] Iryna Sitnikova, "The Ukrainian Satellite 'Sich 2-30' Works in a Limited Mode, It Cannot Transmit Images – the Head of the State Space Agency" ["Український супутник 'Січ 2-30' працює в обмеженому режимі, він не може передавати знімки – голова Держкосмосу"], Hromadske, August 23, 2022.

[13] Michel Laffaiteur, Oleg Fedorov, and Sergey Gerasymchuk, *Space Activities: Ukraine Looking for New Developments*, European Space Policy Institute, August 2011; Government Portal of Ukraine, "Ukraine's Participation in the EU Copernicus Program Will Give Impetus to the Development of Key Sectors of Our State, Assures Ivanna Klympush-Tsintsadze," July 10, 2018.

[14] State Space Agency of Ukraine, "In Ukraine, the Regional Center of the European Union 'Copernicus' Program Has Started Operating" ["В Україні розпочав роботу Регіональний центр програми Європейського Союзу 'Copernicus'"], January 28, 2020; Alla Hurska, "Revival of Ukrainian Space Sector: A Viable New Prospect?" *Eurasia Daily Monitor*, Vol. 17, No. 49, April 13, 2020.

[15] Vsevolod Nekrasov, "A Ukrainian Astronaut on the ISS Is Real. What Will Ukraine's Accession to the European Space Agency Do" ["Український астронавт на МКС – це реально. Що дасть Україні вступ до Європейського космічного агентства"], Ekonomichna Pravda, August 1, 2022.

[16] As defined by U.S. doctrine, the SATCOM mission provides worldwide communications and data transmission, connecting forces around the globe and across the strategic, operational, and tactical levels (Joint Publication 3-14, 2023).

[17] USSF, *United States Space Force Vision for Satellite Communications (SATCOM)*, January 23, 2020, p. 1.

Satellite Communications in Russia

Since the fall of the Soviet Union, Russia has struggled to produce and maintain robust satellite communication capabilities because of low investment, technological challenges, and supply chain issues. The core component of Russia's military SATCOM architecture is the Integrated Satellite Communication System (ESSS), a group of communications satellites composed of two primary constellations—one in highly elliptical orbit and one in geosynchronous equatorial orbit.[18] In total, the ESSS includes roughly 46 satellites, but as many as 30 have reportedly exceeded their warranted lifetime and need to be replaced, limiting the system's battlefield relevance.[19] In addition to dedicated military communications satellites, Russian forces can also access Russia's civilian and commercial SATCOM systems, which include a group of Gonets communications satellites operated by Roscosmos, 14 Express satellites operated by the Russian Satellite Communications Company, and 5 Yamal satellites operated by Gazprom Space Systems.[20] However, the Gonets systems are reportedly outdated as well—suffering from low-bandwidth and high-latency issues—and likely lack sufficient capacity to support Russian warfighting in Ukraine.[21]

The Russian military has sought to replace its aging SATCOM infrastructure, but these efforts have experienced significant delays and were still ongoing when Russia invaded Ukraine in 2022, leaving Russian forces with limited and outdated SATCOM capability.[22] Russia has committed to deploying a third generation of ESSS in highly elliptical orbit and geosynchronous equatorial orbit—called the Sfera-S and Sfera-V, respectively—which also includes a Roscosmos plan to field more than 600 Sfera satellites by 2028. However, progress toward deploying Sfera has been severely hamstrung by supply chain disruptions resulting from Western sanctions imposed on Russia following the invasion.[23] As of October 2022, the Russian military has only launched its first Sfera satellite into orbit, but significant development issues remain unresolved.[24] Therefore, until the Sfera constellation becomes operational, Russia's military systems will continue to lag significantly behind their Western counterparts.[25]

---

[18] Pavel Podvig, "Russian Space Systems and the Risk of Weaponizing Space," in Samuel Bendett, Mathieu Boulègue, Richard Connolly, Margarita Konaev, Pavel Podvig, and Katarzyna Zysk, *Advanced Military Technology in Russia: Capabilities and Implications*, Chatham House, September 2021.

[19] Michael Connell, *The Role of Space in Russia's Operations in Ukraine*, CNA, November 2023.

[20] Connell, 2023; Gazprom Space Systems, "Infrastructure," webpage, undated; Oleksiy Nabozhnyak, "'Russians Have Hundreds of New Satellite Dishes in Kharkiv Oblast' – Radio and Military Technology Consultant," Slidstvo.Info, May 15, 2024.

[21] Pavel Luzin, "Russia's Space Satellite Problems and the War in Ukraine," *Eurasia Daily Monitor*, Vol. 19, No. 76, May 24, 2022; Connell, 2023.

[22] Podvig, 2021.

[23] Roscosmos [Роскосмос], "Assembly and Testing of the First Satellite of the Sphere Project Has Begun" ["Начались сборка и испытания первого спутника проекта 'Сфера'"], August 21, 2022.

[24] "Russia Orbits First Sfera Constellation Satellite," Aviation Week Intelligence Network, October 24, 2022.

[25] Connell, 2023, p. 13.

Satellite Communications in Ukraine

Prior to Russia's 2014 invasion of Crimea, insufficient funding hindered much of Ukraine's military capabilities, including its communications network. Ukrainian soldiers relied on ground lines of communication, both permanent and temporary, which was mostly composed of remnants of Soviet-era equipment.[26] This reliance meant that Ukraine's military networks—including those used by higher-level military officials—were highly susceptible to disruptions from events, such as power outages. Yet, many Ukrainian commanders trained during Soviet or early post-Soviet times still preferred these more-familiar systems to newer ones, which were largely untested by Ukrainian troops.[27]

After the 2014 invasion, newer communication means were introduced—primarily by volunteers—and later integrated into military practice. To provide the troops in Ukraine's eastern territories with critical communications linkages, volunteers started to identify and purchase foreign-made civilian equipment.[28] Among those systems were Tooway terminals, which provided remote consumers access to Viasat's regional satellite internet service.[29] Tooway quickly proved valuable to Ukrainian troops and ultimately became one of the primary means of communication for most battalions and company and platoon strongholds.[30]

By providing a fast and stable internet connection on the frontline and beyond, Tooway enabled more-integrated Ukrainian operations. After 2014, Ukraine developed several internet-enabled software applications that could streamline and improve decisionmaking and combat employment on the battlefield.[31] In particular, Kropyva (Nettle, in English) was originally developed to help Ukrainian armored vehicles, infantry, and reconnaissance units to calculate fire missions and later evolved into a tactical situational awareness system.[32] Delta provides another situational awareness tool that enables coordination among Ukrainian units and offers a common operating picture that draws information from such various sources as drone feeds, social media, and satellite images.[33] To integrate and analyze

---

[26] Oleksandr Lavrut, Tetiana Valeriivna Lavrut, Oleg Klimovich, and Yuriy Mykolaiovych Zdorenko, "New Technologies and Communication Tools in the Armed Forces of Ukraine: Transformation Path and Development Perspectives" ["Новітні технології та засоби зв'язку у ЗСУ: шлях трансформації та перспективи розвитку"], Ukrainian Military Pages, April 17, 2019.

[27] Vitaliy Kuksa, "Two Faces of Military Communication: How American Army Adapts Technology to the Soviet Model" ["Два обличчя військового зв'язку. Як в Армії американську техніку підганяють під радянську модель"], Texty.org.ua, May 5, 2016.

[28] Lavrut et al., 2019.

[29] Tooway is a satellite broadband service provided by the Italian company Skylogic. Tooway operates the consumer-oriented partition of the KA-SAT network owned by the U.S.-based Viasat (Viasat, "KA-SAT Network Cyber Attack Overview," March 30, 2022).

[30] "Special Topic. Military Communication: Transformation" ["Спецтема. Військовий зв'язок: трансформація"], Defense Express, June 7, 2018.

[31] Stefan Soesanto, *The Ukrainian Way of Digital Warfighting: Volunteers, Applications, and Intelligence Sharing Platforms*, Center for Security Studies, July 2024.

[32] Seth G. Jones, Riley McCabe, and Alexander Palmer, *Ukrainian Innovation in a War of Attrition*, Center for Strategic and International Studies, February 2023; Tayisa Melnyk, "Stinging 'Kropyva'. How Ukrainian Software for Artillerymen Affects the Course of the War" ["Жалюча 'Кропива'. Як українське програмне забезпечення для артилеристів впливає на перебіг війни"], *Forbes*, July 24, 2022.

[33] "The War in Ukraine Shows How Technology Is Changing the Battlefield," *The Economist*, July 3, 2023; Soesanto, 2024.

such disparate sources of large amounts of real-time data and quickly communicate information to the frontline, these systems do require internet connectivity. These applications, along with other key internet-enabled software, underscore the importance of SATCOM to provide internet connectivity to deployed forces.[34]

Above and beyond the military, Ukraine's government and broader society were rapidly digitized following the 2014 war, reinforcing the importance of stable internet connectivity for the population.[35] As Russia began mobilizing troops along Ukraine's border in late 2021 and early 2022, many anticipated that Russia would attempt to disrupt Ukraine's internet and mobile communication early on to gain battlefield superiority and sow panic among Ukraine's population.[36] In preparation, Ukraine's government developed handbooks to teach civilians how to stay connected with friends and family without internet and outlined potential courses of action in case of a total disruption of communications.[37]

One possible solution discussed in both military publications and, reportedly, by government officials, was to transition to using Starlink, a commercial internet service provided by SpaceX's proliferated low earth orbit satellite constellation.[38] Although it was not publicized at that time and only became known after the full-scale invasion, then–Deputy Prime Minister and Minister of Digital Transformation of Ukraine Mykhailo Fedorov claimed that, for over a year, his team had been in talks with SpaceX to bring Starlink to Ukraine.[39] Other sources claim that the talks lasted for about a month and a half, with some mentioning a videoconference between chief executive officer of SpaceX Elon Musk and President of Ukraine Volodymyr Zelenskyy to discuss the deliveries and rollout.[40] Further clouding the timeline, a few days before the full-scale invasion, Ukrainian media reported that the start of Starlink service in Ukraine for public use had been postponed from 2022 to 2023.[41] It is unclear how Ukraine planned to proceed with the development of its military communications at that time and, in particular, how it envisioned integrating Starlink, if at all.[42]

---

[34] Vitaliy Manko, "Digital Army of Ukraine: Forewarned Means Armed" ["Цифрова армія України: попереджений – значить озброєний"], Censor.net, December 24, 2019; "How Elon Musk's Satellites Have Saved Ukraine and Changed Warfare," *The Economist*, January 5, 2023; Sam Cranny-Evans, "Have Russian Cyberattacks Changed the Course of the War in Ukraine?" *National Interest*, June 25, 2022.

[35] United Nations Development Programme, "63% of Ukrainians Use State E-Services, User Numbers Grow for Third Year in Row – Survey," press release, January 25, 2023.

[36] David Brown, "Ukraine Conflict: Where Are Russia's Troops?" BBC News, February 23, 2022; James Pearson and Raphael Satter, "Internet in Ukraine Disrupted as Russian Troops Advance," Reuters, February 27, 2022.

[37] Dovidka.Info, "Communication and Information Retrieval," webpage, undated.

[38] Starlink, "High-Speed Internet Around the World," webpage, undated.

[39] Mykhailo Fedorov, "Always Connected: How Starlink by Elon Musk Is Helping Ukraine During the War" ["Завжди на зв'язку. Чим Україні під час війни допомагає Starlink від Ілона Маска"], RBC-Ukraine, June 17, 2022b.

[40] Jeff Foust, "SpaceX Worked for Weeks to Begin Starlink Service in Ukraine," *SpaceNews*, March 8, 2022; Christopher Miller, Mark Scott, and Bryan Bender, "UkraineX: How Elon Musk's Space Satellites Changed the War on the Ground," Politico, June 8, 2022.

[41] Stas Yurasov, "Starlink by Elon Musk Postpones Launch Date of Satellite Internet Service in Ukraine to 2023" ["Starlink Ілона Маска переніс дату запуску послуги супутникового інтернету в Україні на 2023 рік"], dev.ua, February 21, 2022.

[42] Manuals for existing and future military personnel had already considered Starlink as one possible option, but the operational focus remained on Tooway, which was already in use (V. G. Sholudko, M. Yu. Esaulov, O. V. Vakulenko, T. G. Gursky, and M.

The Viasat Hack and the Transition to Starlink

On the morning of February 24, 2022, during the initial hours of Russia's full-scale invasion of Ukraine, thousands of Viasat users—including those using Tooway terminals—were kicked offline by what was later identified as a Russian cyberattack.[43] This attack involved Russian malware that targeted one partition of Viasat's KA-SAT satellite broadband service and succeeded in disconnecting tens of thousands of modems across Ukraine and Europe from the Viasat network.[44] According to a statement from Viasat, Russian hackers exploited a single "misconfiguration in a VPN appliance" to gain access to the system.[45] Viasat further claimed that the network was "largely stabilized within hours" of the attack, although full stabilization took several days. By the time Viasat users were back online, Russian forces had already made significant progress into Ukrainian territory.[46]

Although the Russian attack on Viasat successfully disrupted Ukrainian internet services, its ultimate impact on Ukrainian military operations remains uncertain. Initially, Ukrainian cybersecurity expert Victor Zhora described the attack as causing "a really huge loss in communications in the very beginning of the war."[47] However, in a later interview, Zhora again confirmed that the attack "resulted in outage of [SATCOM]," but then clarified that, nevertheless, "it didn't impact the process of coordination between forces and between state leaders and forces."[48] This was attributed to the Ukrainian military's continued reliance on ground lines rather than satellites for much of its communication. Lt Gen Jack Weinstein, USAF (Ret.), echoed Zhora's remarks, noting that the Ukrainian military has "built in resilient communications to make sure that they were able to quickly alter measures and keep fighting."[49] Despite the Ukrainian military's ability to adjust to such an incident, the Viasat hack highlighted the need for commercial firms to improve their awareness of and defenses against cyberattacks.

The Viasat hack exposed the vulnerabilities of relying on Tooway for both civilian and military purposes, prompting Ukraine to pursue more-resilient communication capabilities.[50] Among the leading candidates was Starlink, which was already deemed capable of supporting communications among Ukrainian civilians and forces. Shortly after the onset of Russia's full-scale invasion, Minister Fedorov famously posted to Twitter, urging Musk to "provide Ukraine with Starlink stations and to

---

M. Fomin, *Organization of Military Communication: Training Manual* [*Організація військового зв'язку: навчальник посібник*], Ministry of Defence of Ukraine, 2017.

[43] Viasat, 2022.

[44] Viasat, 2022.

[45] Viasat, 2022.

[46] Viasat, 2022.

[47] Raphael Satter, "Satellite Outage Caused 'Huge Loss in Communications' at War's Outset – Ukrainian Official," Reuters, March 15, 2022.

[48] Kim Zetter, "Viasat Hack 'Did Not' Have Huge Impact on Ukrainian Military Communications, Official Says," Zero Day, September 26, 2022.

[49] Jonathan Greig, "CISA's Goldstein: Ukrainian Response to Viasat Hack Proves Need for Redundancy, Resilience," The Record, November 15, 2023.

[50] Cranny-Evans, 2022.

address sane Russians to stand."[51] Although friction between Ukraine and Musk would develop later, Musk initially agreed. On February 26, 2022, SpaceX activated Starlink, and, a few days later, terminals were delivered to Ukraine.[52] As early as February 28, 2022, Minister Fedorov posted a photo to the X platform showing a truckload of Starlink terminals arriving in Ukraine.[53] Within a week after the start of the full-scale invasion, the majority of the 500 terminals sent to Ukraine became operational.[54] Although this number of terminals was unlikely to restore internet access for the entire country in the event of a much-feared nationwide blackout, it provided hot spots for essential operations, such as those for government or journalists.[55] In the first few weeks following Starlink's arrival, Ukraine registered approximately 150,000 daily users of the service.[56] The monthly fee was also waived.[57]

Several other initiatives were undertaken to extend Starlink services to a variety of users in Ukraine. By the first week of April 2022, the U.S. Agency for International Development provided 5,000 terminals to Ukraine through a public-private partnership.[58] The agency funded 1,330 of these terminals, and SpaceX donated an additional 3,670 terminals and supplied the internet service.[59] Later in the war, Germany allocated €20 million for Ukraine to acquire additional Starlink terminals and services.[60]

Ukraine's early access to Starlink proved particularly important during the initial weeks of the full-scale invasion, a period marked by high uncertainty when the ability to get information became critical to avoiding panic among the population.[61] According to the Ukrainian Ministry of Health, within the first month of service provision, 590 medical facilities were equipped with Starlink, enabling satellite internet connectivity even in areas of active combat. The terminals were also distributed to local

---

[51] Mykhailo Fedorov [@FedorovMykhailo], "@elonmusk, while you try to colonize Mars — Russia try to occupy Ukraine! While your rockets successfully land from space — Russian rockets attack Ukrainian civil people! We ask you to provide Ukraine with Starlink stations and to address sane Russians to stand." post on the X platform, February 26, 2022a. At the time of Minister Fedorov's post, the platform was still known as Twitter; Musk had not yet acquired Twitter and had not yet changed its name to X.

[52] Foust, 2022.

[53] Chelsea Gohd, "SpaceX Starlink Satellite Internet Terminals Arrive in Ukraine," Space.com, February 28, 2022.

[54] Miller, Scott, and Bender, 2022.

[55] Jackie Wattles, "SpaceX Sent Starlink Internet Terminals to Ukraine. They Could Paint a 'Giant Target' on Users' Backs, Experts Say," CNN, March 4, 2022.

[56] Kate Duffy, "SpaceX Starlink Has 150,000 Daily Users in Ukraine 5 Weeks After Being Activated, Government Official Says," Business Insider, May 3, 2022.

[57] Daryna Antoniuk, "How Elon Musk's Starlink Satellite Internet Keeps Ukraine Online," *Kyiv Independent*, September 3, 2022.

[58] U.S. Agency for International Development, "USAID Safeguards Internet Access in Ukraine Through Public-Private-Partnership with SpaceX," press release, April 5, 2022; Miller, Scott, and Bender, 2022; Olga Boichak and Tetyana Lokot, "Billionaires Won't Save Ukraine's Internet," *Foreign Policy*, November 20, 2022.

[59] France confirmed that it also helped in this effort and that the systems were transported through Poland (Cristiano Lima-Strong, "U.S. Quietly Paying Millions to Send Starlink Terminals to Ukraine, Contrary to SpaceX Claims," *Washington Post*, April 8, 2022).

[60] Oleksiy Yarmolenko, "Germany Will Allocate €20 Million to Pay for 'Starlink' Services in Ukraine," Babel, January 10, 2023.

[61] Miller, Scott, and Bender, 2022.

communities, schools, government service centers, mass media, volunteer centers, and critical infrastructure, such as trains used primarily for civilian evacuation.[62]

As the war continued, Starlink played a key role in ensuring that government and business entities remained operational.[63] When Ukraine started achieving battlefield successes, the terminals facilitated the rapid restoration of mobile and internet communications in de-occupied cities, including Bucha or Hostomel.[64] In fall 2022, as Russia intensified its attacks on Ukraine's energy infrastructure, Starlink terminals were used to restore cell phone services and network connectivity in the areas affected by blackouts.[65] This connectivity was instrumental in keeping the global community informed on situations on the battlefield and atrocities committed by Russian troops.[66] Ukrainian civilians used designated channels and applications to report sightings of Russian military equipment or personnel, providing almost instantaneous alerts to Ukrainian forces. Civilians also shared their experiences under occupation and showed defiance through the creation of numerous memes and viral videos.[67]

In addition to various civilian applications, Starlink became instrumental for such military command and control systems as Kropyva and Delta, which had previously relied on Tooway terminals. Starlink's terminals were relatively easy to set up, becoming operational in under 15 minutes.[68] This ease of use, combined with the functionality of Ukraine's situational awareness systems, proved particularly valuable, allowing Ukrainian units to receive orders, access military manuals, or obtain online support.[69] Beyond command and control systems, Starlink facilitated Ukrainian drone warfare, intelligence collection and fusion, and fire support operations.[70] Ukraine established command posts that monitored the combat environment online using unmanned aerial vehicle (UAV) cameras.[71] In the first month of the war alone, the Ukrainian drone unit Aerorozvidka performed about 300 information gathering missions daily.[72] The unit used Starlink terminals to transmit the collected data to artillery teams, who relied on this intelligence to fire at enemy units and

[62] Fedorov, 2022b; Kurt Vinion, "How Elon Musk's Starlink Became Invaluable to Ukraine's War Effort," Radio Free Europe/Radio Liberty, October 20, 2022; Violetta Orlova, "The Ministry of Digital Transformation of Ukraine Explained How Elon Musk's Starlink Internet Will Help Ukrainians in the War" ["У Мінцифри розповіли, як інтернет Ілона Маска Starlink допоможе українцям на війні"], Unian, March 7, 2022; Denys Katsylo, "We Go Wherever We Can. The Head of the Board of 'Ukrzaliznytsia' Talks About Work During the War and Elon Musk's Starlink" ["Їдемо всюди, де тільки можемо. Голова правління 'Укрзалізниці' – про роботу під час війни та Starlink від Ілона Маска"], *Forbes*, March 2, 2022.

[63] U.S. Agency for International Development, 2022.

[64] "Lifecell Partially Restored Mobile Communication in Bucha Using Starlink" ["Lifecell частково відновив у Бучі мобільний зв'язок за допомогою Starlink"], Ukrinform, April 8, 2022; Fedorov, 2022b.

[65] Mohar Chatterjee, "Ukraine Scrambles to Keep Internet Up amid Blackouts," Politico, October 21, 2022.

[66] Alex Horton and Serhii Korolchuk, "Whatever the Fuss over Elon Musk, Starlink Is Utterly Essential in Ukraine," *Washington Post*, September 18, 2023.

[67] Steven Feldstein, "Disentangling the Digital Battlefield: How the Internet Has Changed War," *War on the Rocks*, December 7, 2022.

[68] "Russia Using Thousands of SpaceX Starlink Terminals in Ukraine, WSJ Says," Reuters, February 15, 2024.

[69] Horton and Korolchuk, 2023; "Starlink's Performance in Ukraine Has Ignited a New Space Race," *The Economist*, January 5, 2023.

[70] Jones, McCabe, and Palmer, 2023.

[71] "SpaceX Relinquishes Control of 'Ukrainian' Starlink Terminals to Pentagon," Militarnyi, September 14, 2023.

[72] Alexander Freund, "Ukraine Using Starlink for Drone Strikes," Deutsche Welle, March 27, 2022.

military systems with greater precision.[73] A network of UAVs connected through Starlink helped Ukraine to successfully deploy antitank weapons in various battlefield conditions, including at night and with minimal noise emissions.[74] The ability to connect reconnaissance equipment and artillery units reduced the time from target detection to fire engagement to only 30 minutes or less.[75] Starlink's impact was so significant that some Ukrainian commanders believed that, without it, the war could have been lost.[76]

Recognizing the importance of Starlink in Ukrainian military operations, in October 2022, Russia warned that Western commercial satellites could become "legitimate targets" in the war.[77] In a statement made during a United Nations meeting, senior Russian foreign ministry official Konstantin Vorontsov characterized the Western use of commercial satellites in Ukraine as "provocative" and described it as "an extremely dangerous trend that goes beyond the harmless use of outer space technologies."[78] He further cautioned the United Nations that employing these "quasi-civilian" satellites risked escalating the ongoing space arms race.[79] This statement prompted concerns that Russian attacks on commercial systems might expand beyond electronic and other more-reversible counterspace threats to include the use of kinetic, debris-producing weapons.[80] The United States responded to these Russian threats by noting Russia's own contributions to the space arms race, specifically referencing the 2021 Russian direct-ascent anti-satellite test and emphasizing the dangers associated with kinetic weapon tests.[81] In October 2022, after Russian officials at the United Nations reiterated their threats to Western commercial systems, White House officials did not specify potential consequences for such actions but did note that "any attack on U.S. infrastructure will be met with a response . . . in a time and manner of our choosing."[82]

As of this writing in 2025, reports of actual Russian attempts to target Western commercial satellites have been relatively mild and exclusively composed of nonkinetic attacks, largely with limited effects.[83] In the early stages of the full-scale invasion, the Russian military attempted to disrupt Starlink through a series of hacking and jamming efforts. SpaceX successfully repelled these attacks by

---

[73] Charlie Parker, "Specialist Ukrainian Drone Unit Picks Off Invading Russian Forces as They Sleep," *The Times*, March 18, 2022.

[74] Vinion, 2022.

[75] "SpaceX Relinquishes Control of 'Ukrainian' Starlink Terminals to Pentagon," 2023; Victoria Kim, Richard Pérez-Peña, and Andrew E. Kramer, "Elon Musk Refused to Enable Ukraine Drone Attack on Russian Fleet," *New York Times*, September 8, 2023.

[76] Yaroslav Trofimov, Micah Maidenberg, and Drew FitzGerald, "Ukraine Leans on Elon Musk's Starlink in Fight Against Russia," *Wall Street Journal*, July 16, 2022.

[77] "Russia Warns West: We Can Target Your Commercial Satellites," Reuters, October 27, 2022.

[78] Brett Tingley, "Russia Says Private Satellites Could Become 'Legitimate Target' During Wartime," Space.com, September 16, 2022a.

[79] Brandon Vigliarolo, "Russia Says Starlink Satellites Could Become Military Target," *The Register*, October 28, 2022.

[80] Isabel van Brugen, "Russia Threatens to Shoot Down Western Satellites," *Newsweek*, October 19, 2023.

[81] Vigliarolo, 2022.

[82] Brett Tingley, "White House Says US Would Respond If Russia Targets Commercial Satellites," Space.com, October 28, 2022b.

[83] Kari A. Bingen, Kaitlyn Johnson, and Makena Young, *Space Threat Assessment 2023*, Center for Strategic and International Studies, April 2023; "Russia Warns West: We Can Target Your Commercial Satellites," 2022.

implementing a rapid update to its systems and minimized interference with or damage to Starlink systems.[84]

Starlink's Dependability Concerns

The foremost concerns about Starlink have stemmed not from direct Russian interference but from uncertainties around SpaceX's reliability as a provider. Because SpaceX supplied Ukraine with Starlink coverage voluntarily using an informal agreement struck on X, formerly known as Twitter, the company was not beholden to any contractual terms or subject to dedicated government oversight.[85] As a result, SpaceX covered most of the associated provision costs—reportedly $80 million over six months—and decided when, where, and how Starlink would be provided and used. This arrangement—or lack thereof—gave rise to multiple instances in which SpaceX's company policy and statements made by Musk conflicted with Ukrainian priorities, raising questions about Starlink's dependability.[86]

One major sticking point involved SpaceX's reservations over the direct use of Starlink in Ukrainian offensive operations. SpaceX offers two services: Starlink, which its website indicates "is designed for consumer and commercial use," and Starshield, which is "designed for government use."[87] SpaceX seemed to envision Starlink being used for humanitarian purposes, for connecting Ukrainian soldiers, and perhaps for limited military functions, such as planning operations.[88] Ukrainian forces, however, increasingly adapted Starlink for military purposes, including for communication among fielded units and for connecting with drones.[89] Tension between Ukraine and SpaceX soon emerged. For example, as Ukrainian soldiers liberated occupied territories along the frontlines in Kherson, Zaporizhzhia, Kharkiv, Donetsk, and Luhansk, their Starlink terminals began experiencing connection issues.[90] Media reports suggested that the outages occurred because SpaceX was geofencing its Starlink service and had not yet activated its service in these newly reclaimed

---

[84] Eric Mack, "US Military Says SpaceX Handily Fought Off Russian Starlink Jamming Attempts," CNET, April 22, 2022; Bingen, Johnson, and Young, 2023, p. 17.

[85] Foust, 2022.

[86] Sebastian Tong, "Musk Says SpaceX Has Spent $80 Million Out of Pocket to Support Ukraine," Bloomberg, October 3, 2022; Tara Copp, "Elon Musk's Refusal to Have Starlink Support Ukraine Attack in Crimea Raises Questions for Pentagon," Associated Press, September 11, 2023.

[87] SpaceX, "Starshield," webpage, undated; Starlink, undated.

[88] Jon Porter, "SpaceX Inks First Space Force Deal for Government-Focused Starshield Satellite Network," The Verge, September 28, 2023.

[89] Ukrainian troops were also reported to have integrated Starlink terminals directly into the fuselages of UAVs with varying ranges and costs to achieve diverse objectives from basic observation to night targeting, which requires a much higher precision. See Joey Roulette, "SpaceX Curbed Ukraine's Use of Starlink Internet for Drones – Company President," Reuters, February 9, 2023; Sam Skove, "How Elon Musk's Starlink Is Still Helping Ukraine's Defenders," Defense One, March 1, 2023a.

[90] Mehul Srivastava, Roman Olearchyk, Felicia Schwartz, and Christopher Miller, "Ukrainian Forces Report Starlink Outages During Push Against Russia," *Financial Times*, October 7, 2022.

territories.[91] At that time, Minister Fedorov did give Starlink credit for "very promptly" responding to Ukraine's concerns in that instance. [92]

In September 2022, Ukraine was unable to execute a naval attack because of a lack of Starlink service in Crimea, although this event was only publicly reported following the 2023 publication of a biography of Musk.[93] The Ukrainian government requested an emergency exemption from the geofencing policy to secure Starlink connectivity for planned naval and drone attacks in the area.[94] However, Musk denied the request, arguing that these attacks could potentially escalate the war beyond Ukraine's borders.[95] Ukrainian officials rebuked the decision as having devastating consequences for Ukraine and further enabling Russia to target Ukrainian cities.[96] U.S. military officials declined to comment on Musk's decision directly, stating only that "[t]he Department continues to work closely with commercial industry to ensure we have the right capabilities the Ukrainians need to defend themselves."[97]

These tensions came to a head in October 2022 when Musk publicly threatened to terminate Ukraine's Starlink services entirely, citing financial fatigue at the prospect of SpaceX providing indefinite free coverage in Ukraine.[98] This threat came just after Musk had made several controversial public statements—including a post to the X platform suggesting that Ukraine should consider surrendering Crimea and other territories in a potential peace agreement—which drew immense criticism from Ukrainian officials.[99] A month prior, SpaceX had privately expressed the same financial concerns in a letter to the Pentagon and outright requested that the U.S. Department of Defense begin footing the bill.[100] As Ukraine and its Western allies feared that Musk had soured on the handshake arrangement and that this now crucial warfighting capability could soon be disabled, the United States entered negotiations with Musk and SpaceX to formally contract the provision of

---

[91] Adam Satariano, Scott Reinhard, Cade Metz, Sheera Frenkel, and Malika Khurana, "Elon Musk's Unmatched Power in the Stars," *New York Times*, July 28, 2023; Ronan Farrow, "Elon Musk's Shadow Rule," *New Yorker*, August 21, 2023.

[92] Satariano et al., 2023.

[93] The book originally indicated that SpaceX had turned off the coverage, although it was later reported that it had never been turned on (Dearbail Jordan, "Elon Musk Says He Withheld Starlink over Crimea to Avoid Escalation," BBC News, September 8, 2023).

[94] Satariano et al., 2023.

[95] "Musk Says He Refused Kyiv Request for Starlink Use in Attack on Russia," Reuters, September 8, 2023; Elon Musk [@elonmusk], "There was an emergency request from government authorities to activate Starlink all the way to Sevastopol. The obvious intent being to sink most of the Russian fleet at anchor. If I had agreed to their request, then SpaceX would be explicitly complicit in a major act of war and conflict escalation." post on the X platform, September 7, 2023.

[96] "Elon Musk Does Not Consider Himself a Traitor, 'Because Congress Did Not Declare War on Russia'" ["Ілон Маск не вважає себе зрадником, 'бо Конгрес не оголошував війну Росії'"], Voice of America, September 11, 2023.

[97] "Musk Says He Refused Kyiv Request for Starlink Use in Attack on Russia," 2023.

[98] Isabelle Khurshudyan, Kostiantyn Khudov, Dan Lamothe, and Ellen Francis, "Musk Threatens to Stop Funding Starlink Internet Ukraine Relies on in War," *Washington Post*, October 14, 2022.

[99] Joseph Wilson, "Musk's Plan to End Russian War Infuriates Ukraine on Twitter," Associated Press, October 4, 2022.

[100] Alex Marquardt, "Exclusive: Musk's SpaceX Says It Can no Longer Pay for Critical Satellite Services in Ukraine, Asks Pentagon to Pick Up the Tab," CNN, October 14, 2022.

Starlink in Ukraine.[101] However, this required convincing Musk to continue providing coverage until the negotiations were finalized. It was later revealed that this process might have been greatly complicated by reported direct communications between Musk and President Putin. Nevertheless, Musk eventually promised to keep Ukraine's Starlink services operational throughout the negotiating term.[102] This arrangement lasted until June 2023, when DoD announced that SpaceX had been awarded a formal contract to provide Starlink coverage in Ukraine.[103] The contract—renewed through 2024—not only ensured the continued provision of Starlink but also established contractual terms that more closely aligned Starlink's operational practices with U.S. national objectives.[104]

In February 2024, new reports claimed that Russia has been using thousands of Starlink terminals in its operations against Ukraine, raising fresh concerns about Starlink's involvement in the war.[105] This usage has reportedly not only enabled Russian attacks but also possibly degraded the quality of connections for Ukrainian users.[106] Musk denied that any Starlink terminals were sold directly to Russia, and Ukrainian officials have suggested that Russia is obtaining the terminals from third-party suppliers and capturing them on the battlefield.[107] This prompted some U.S. lawmakers to question Musk's adherence to U.S. sanctions, as they wrote in a letter: "We are concerned that you may not have appropriate guardrails and policies [. . .] to ensure [SpaceX's] technology is neither acquired directly or indirectly, nor used illegally by Russia."[108] Starlink is not the only U.S. supplier facing challenges in limiting Russia's use of their technology. For example, there have also been reports of Russia taking advantage of commercial imagery intended to assist Ukraine in collecting information on targets within Ukraine.[109]

In April 2024, DoD and the USSF separately released strategy documents on incorporating commercial space capabilities, which, to some degree, acknowledged challenges and opportunities presented by the role of Starlink in the Russia-Ukraine war. For example, the documents recognized the contribution of commercial capabilities for SATCOM, emphasized the need for integration *prior* to the crisis, and the potential for "the use of military force to protect and defend commercial assets" under "appropriate circumstances."[110] These policy documents also called out demands for financial

---

[101] Farrow, 2023; Tara Copp and Lolita C. Baldor, "Officials: Musk Seeks US Funds for Ukraine Satellite Network," Associated Press, October 15, 2022; Cristian Segura, "Ukraine Places Trust in White House to Prevent Musk Pulling Plug on Starlink Comms for the Armed Forces," *El País*, September 19, 2023.

[102] "Starlink: Why Elon Musk's Internet Is so Important for Ukraine" ["Starlink: чому інтернет від Ілона Маска такий важливий для України"], BBC News Ukraine, October 19, 2022; Farrow, 2023.

[103] Mike Stone and Joey Roulette, "SpaceX's Starlink Wins Pentagon Contract for Satellite Services to Ukraine," Reuters, June 1, 2023.

[104] Anthony Capaccio, "Pentagon Deal with Musk's Starlink in Ukraine Extended Six Months for $14 Million," Bloomberg, June 13, 2024.

[105] James Marson and Thomas Grove, "Russia Using Thousands of Musk's Starlink Systems in War, Ukrainian General Says," *Wall Street Journal*, February 15, 2024.

[106] Marson and Grove, 2024.

[107] Mike Wendling, "US Lawmakers Quiz Musk's Starlink over Russia Claims," BBC News, March 7, 2024.

[108] Wendling, 2024.

[109] Graeme Wood, "A Suspicious Pattern Alarming the Ukrainian Military," *The Atlantic*, March 18, 2024; Michael Marrow, "Pentagon Working with Ukraine, SpaceX to Prevent Russian Exploitation of Starlink," Breaking Defense, April 10, 2024.

[110] DoD, *Commercial Space Integration Strategy*, 2024, p. 3; USSF, *U.S. Space Force Commercial Space Strategy: Accelerating the Purposeful Pursuit of Hybrid Space Architectures*, April 8, 2024.

protection and a variety of contracting tools to leverage commercial capabilities. DoD policy statements on commercial space include the idea of establishing norms to deter adversary attacks, while other analysts have proposed using compliance with United Nations guidelines to discourage malign behavior.[111]

However, not all questions raised by the use of Starlink were fully answered in these strategy documents, including how or when DoD might respond to attacks on commercial satellites, the completeness of existing contractual arrangements, or the extent to which U.S. companies can guarantee reliable support for U.S. partner or allied countries in crisis or conflict. For example, Taiwan reportedly is cautious about relying on SpaceX as a military provider, given Musk's close ties with China as the chief executive officer of Tesla, which produces approximately 50 percent of its new cars in China.[112]

Key Takeaways

By the time of Russia's full-scale invasion of Ukraine, years of undercapitalization and neglect had left Russian forces with an outdated and insufficient SATCOM capability, ostensibly limiting the role of satellite connectivity in Russia's warfighting, particularly in the conflict's early stages. For Ukraine, however, SATCOM services—most notably those provided by commercial vendors—have proven indispensable throughout the war, aided by pre-invasion investments in integrated operations and web-based applications for situational awareness and mission planning. SpaceX's voluntary provision of Starlink services filled gaps created by Russia's Viasat hack and enabled Ukrainian intelligence fusion, fire control, situational awareness, drone operations, and real-time tactical command and control. By one account, by May 2024, Ukrainian forces were using 60,000–70,000 Starlink terminals on the front line, demonstrating the continuing ubiquity and importance of the system.[113]

However, the informal nature of SpaceX's initial commitment allowed multiple flashpoints to emerge between company leadership priorities and Ukraine's warfighting objectives, casting doubt on the reliability of Starlink until the Pentagon officially contracted its provision. Musk's particular opinions and control over SpaceX played an outsized role at some key junctures of the war, which points to the importance of powerful individuals in the commercial space industry under particular circumstances. However, the provision of Starlink to Ukraine without a contract was likely a unique circumstance. Nevertheless, these events have heightened the importance of developing formal and enforceable agreements with future commercial space partners. Although DoD documents reflect an awareness of this challenge, there remain questions about the extent of DoD protection of commercial assets in conflict and the reliability of commercial providers for countries in conflict.

---

[111] DoD, 2024, p. 4; Bruce McClintock and Mélusine Lebret, "Russian Space Strategy and Capabilities: A Tale of Decline," in Alessandro Gili, ed., *The Sky Is Not the Limit: Geopolitics and Economics of the New Space Race*, Ledizioni LediPublishing, 2024, pp. 142–143.

[112] Satariano et al., 2023.

[113] Nabozhnyak, 2024.

Positioning, Navigation, and Timing

PNT refers to the provision of precise positioning, navigation, and time reference services for use across various applications.[114] Although PNT can be locally provided or augmented with terrestrial assets, space-based constellations have been used as the main tools to deliver continuous PNT signals anywhere on earth. Such global navigation satellite systems (GNSS) have commonly been placed in medium earth orbit, which implies the need for about 18–30 satellites for global coverage.[115] As of this writing in 2025, there are four such GNSS in operation: the U.S. Global Positioning System (GPS), Russia's Global Navigation Satellite System (GLONASS), China's BeiDou, and the European Union's Galileo.[116] These systems broadcast a civilian signal available to commonly produced commercial receivers used in consumer electronics, as well as signals used on military-dedicated frequencies, which are generally more precise but require specialized receivers. Although the GPS-aided munitions eventually provided to Ukraine use GPS signals, other weapons producers could make use of commercial receivers that incorporate GPS and possibly other GNSS services.[117] This section discusses the role of space-based PNT in the war in Ukraine, with a particular focus on GNSS-guided weapons and efforts from both Russia and Ukraine to interfere with satellite PNT signals.

Russia's GNSS-Enabled Attacks

For Russia, GNSS-guided munitions played a limited role in the early days of fighting in Ukraine, but reports have shown increased Russian success in employing such satellite-guided weapons as glide bombs and Shahed-type drones.[118] Historically, the Soviet Union neglected the development of conventional guided weapons until the mid-1980s, and Russia only began seriously pursuing their development in the past two decades.[119] During that period, Russia was able to field a modest arsenal of precision-guided munitions (PGMs) that were equipped with inertial guidance systems that were then supplemented with military-dedicated GLONASS (and sometimes commercial GPS) receivers to achieve even greater strike precision. However, while developing these weapons, Russia routinely underinvested in the upkeep of GLONASS itself, leading to reported deterioration and significant reliability issues.[120] Although the constellation was restored to operational condition in the early 2010s and formally accepted into Russia's military service in 2016, many of its satellites soon neared the end

---

[114] Joint Publication 3-14, 2023.

[115] Advanced Navigation, "Global Navigation Satellite System (GNSS) and Satellite Navigation Explained," March 8, 2023.

[116] GPS.gov, "Other Global Navigation Satellite Systems (GNSS)," webpage, October 19, 2021.

[117] For example, the Guided Multiple Launch Rocket Systems (GMLRS) rocket—one of several guided munitions provided to Ukraine—is aided by GPS (Lockheed Martin, "US Army Awards $4.79 Billion Production Contract for All-Weather GMLRS Rocket," press release, April 27, 2023; Connell, 2023, p. 14).

[118] Giorgio Di Mizio and Douglas Barrie, "Russian Glide Bombs Add Pressure on Ukraine's Diminishing Air Defences," International Institute for Strategic Studies, March 25, 2024; Uzi Rubin, "Russia's Iranian-Made UAVs: A Technical Profile," Royal United Services Institute for Defence and Security Studies, January 13, 2023.

[119] John Ismay, "Russian Guided Weapons Miss the Mark, U.S. Defense Officials Say," *New York Times*, May 9, 2022a.

[120] Podvig, 2021.

of their lifespans.[121] Thus, by the time Russia began its full-scale invasion of Ukraine in 2022, its arsenal of PGMs was largely limited, untested, and reliant on a degraded GNSS system for critical accuracy improvements.

Still, the early stages of the war saw Russia launching precision-guided strikes against Ukrainian targets. These attacks were often unsuccessful, with missiles reportedly missing their targets or even failing to work entirely.[122] As time progressed, Russia encountered issues sustaining both its guided weapons program and GLONASS, particularly in the face of Western sanctions. After exhausting much of its PGM stockpile early in the war—and to limited effect—trade restrictions hamstrung Russian efforts to acquire more PGMs, leaving their forces at that time with predominantly "dumb" unguided bombs.[123] The sanctions have also limited Russia's options for sourcing the space-grade, radiation-resistant electronics required for new GLONASS satellites, potentially limiting the accuracy of any GNSS-guided munitions Russia could have produced.[124]

However, despite these initial challenges, the Russian military continued its development of PGMs and its employment of other PNT-enabled weapons; more-recent episodes suggest that its guided weapons and space programs might be improving.[125] For example, in February 2024, Russia launched the hypersonic Zircon missile for the first time in two years of war.[126] Although its target was unclear and details about its guidance system remain unknown, Zircon successfully reached Ukrainian territory and heavily damaged a site in Kyiv.[127] Beyond such exquisite systems, reports of Russia's extensive use of and success with satellite-guided glide bombs, Shahed-type drones, cruise missiles, and other long-range weapons to strike Ukrainian infrastructure might also indicate greater use of precision targeting. These munitions can use a variety of different systems for targeting, including inertial guidance, GLONASS-enabled systems, or commercial GNSS receivers.[128] Improvements to GLONASS since the start of the full-scale war—such as the launches of its 24th and final GLONASS-M satellite in 2022 and its first fourth-generation GLONASS-K2 satellite in 2023—might allow Russia to rely more on its native systems.[129] Nevertheless, its ability to conduct long-range strikes despite the weakness of the GLONASS constellation highlights how adversaries can depend on multiple PNT systems to effectively conduct operations.

---

[121] See McClintock and Lebret, 2024.

[122] Ismay, 2022a.

[123] Ismay, 2022a.

[124] Connell, 2023.

[125] Oleksandr Burlaka, "What Russian Satellites See," Universe Space Tech, November 9, 2023.

[126] Brad Lendon, "Russia Used an Advanced Hypersonic Missile for the First Time in Recent Strike, Ukraine Claims," CNN, February 13, 2024.

[127] "Russia Uses Zircon Hypersonic Missile in Ukraine for First Time, Researchers Say," Reuters, February 12, 2024.

[128] By one account, different Shahed drones relied on inertial navigation, a commercial GPS system, a GLONASS-enabled system, and the Ukrainian cellular telephone network for guidance ("A Technophile's Guide to the Evolution of Russian Shahed Drones," *Kyiv Post*, December 10, 2023).

[129] Mike Wall, "Russia Launches Final GLONASS-M Navigation Satellite into Orbit," Space.com, November 30, 2022.

Ukraine's GNSS-Enabled Attacks

Ukrainian efforts to develop guided missile technology began as early as 1994.[130] However, domestic production issues, including lack of funding, limited access to technology, corruption, and long timelines hampered its pursuit of such expensive systems.[131] Even at the time of Russia's initial invasion in 2014, Ukraine's primary military objectives were to resist Russian-aided separatists and to defend its eastern territories from imminent Russian threats, such as enemy fighters entering Ukrainian airspace. These priorities took precedence over—and might not have immediately benefited from—developing PGMs.[132] However, in the years following, Ukraine began to actively reform its military and modernize its equipment, including renewed efforts to build indigenous GPS-aided munitions. In January 2016, the National Security and Defense Council of Ukraine launched a program to develop and domestically produce high-precision weapons systems.[133] The guided missile named Vilkha underwent multiple successful tests in 2016–2018 and was ultimately fielded by the Ukrainian military.[134] In April 2020, Ukraine reported that a modernized Vilkha variant, Vilkha-M, was capable of hitting a 15-m target located 110 km away.[135] To achieve such precision, the system uses GPS-aided inertial guidance and 90 guidance thrusters.[136]

These capabilities proved immediately useful when Russia waged its full-scale invasion in February 2022. Within the war's first weeks, Ukraine began launching Vilkha missiles against Russian forces. By March 2022, the State Kyiv Luch Design Bureau—Vilkha's manufacturer—published a (since deleted) post claiming 50 successful strikes with these PGMs.[137] However, this usage rate quickly proved unsustainable, as limited pre-invasion stockpiles and longstanding production issues constrained the number of Vilkha systems Ukraine could field.[138] In February 2023, a representative of Ukraine's defense industrial base acknowledged that production of Vilkha had resumed. Subsequently, additional reports have surfaced of Vilkha systems striking Russian positions far from

[130] Volodymyr B., "Vilkha: Ghost Weapon or Game Changer?" ["Вільха: зброя-привид чи gamechanger?"], Militarnyi, October 23, 2023; "Ukrainian Tactical Missiles. Instead of Designing Until 2020, It's Better to Buy Them from the US Right Now" ["Українські тактичні ракети. Замість проектувати до 2020 року – краще їх купити у США прямо зараз"], Texty.org.ua, January 22, 2010.

[131] Serhii Dobrynin and Yevhen Lehalov, "'Relative' of HIMARS: What Can the Ukrainian Missile Complex 'Vilkha' Do" ["'Родич' HIMARS: на що здатний український ракетний комплекс 'Вільха'"], Radio Liberty, January 15, 2024; "Ukrainian Tactical Missiles: Instead of Designing Until 2020, It's Better to Buy Them from the US Right Now," 2010.

[132] Olena Maksymenko, "General Director of Luch Design Bureau Oleh Korostelov: 'The Military Requests 2,000 Missiles, but We Can Only Provide 600 or 800'" ["Гендиректор КБ 'Луч' Олег Коростельов: 'Військові пишуть, що їм потрібно 2000 ракет, а ми їм забезпечуємо тільки 600 або 800'"], Texty.org.ua, September 6, 2021.

[133] State Kyiv Design Bureau "Luch," homepage, undated.

[134] The Vilkha system has several variations: R624-M Vilkha-M (a range of 130 km and a payload of 170 kg), R624-M1 Vilkha-M1 (variant 1 with a range of 154 km and a payload of 170 kg; variant 2 with a range of 121 km and a payload of 236 kg), R624-M2 Vilkha-M2 (the range can vary from 141 to 202 km and a payload of 170 kg), R624-P Vilkha-P (with cluster munitions) (B., 2023).

[135] "Vilkha-M Hit a Target at 110 Km with a Deviation of 15 M" ["Вільха-М вразила ціль на 110 км з відхиленням у 15 м"], Ukrainian Military Pages, April 26, 2020.

[136] Stijn Mitzer and Joost Oliemans, "Novel Capabilities: Ukraine's Vilkha Guided MRL," Oryx, May 10, 2021.

[137] "Ukraine Launched a Massive Strike Using Rare Vilkha Rockets, Russia Claims," Defense Express, January 3, 2024.

[138] "Ukraine Launched a Massive Strike Using Rare Vilkha Rockets, Russia Claims," 2024.

the front line and even inside Russia, despite repeated claims from the Russian military that said attacks were successfully intercepted.[139]

However, even with the eventual resumption of Vilkha production, Ukraine's indigenous PGM capacity was never sufficient for a large-scale protracted war with Russia. Instead, Ukraine has looked to foreign military assistance—particularly from the United States—to provide the lion's share of its precision-guided weapons, among other capabilities. The need for these military systems already became apparent in 2014 following Russia's annexation of Crimea, prompting Ukraine to seek Western military aid.[140] The United States and its allies agreed to provide assistance but focused their support on nonlethal, defensive capabilities.[141] This restriction then loosened somewhat under U.S. President Donald Trump's first term, which sold Ukraine over 200 Javelin anti-tank missiles and provided Ukraine with miscellaneous small arms.[142] However, even in the wake of Russia's full-scale invasion, the United States was initially hesitant to provide heavier and longer-range systems.[143] It was only a few months into the war when the White House began sending Ukraine increasingly capable GPS-guided weapon systems, as shown in Table 1.

When Ukraine received its first complete package of GPS-guided weapons—GMLRS missiles fired from HIMARS launchers—it was heralded as a game-changer on the battlefield.[144] These and subsequent PGM systems enabled Ukraine to engage with a significantly wider variety of high-value targets, such as in attacks on the Antonivsky Bridge and Belbek airfield.[145] However, the United States has consistently imposed restrictions on what Ukraine can target with these munitions, most notably forbidding strikes inside Russian territory—though this was narrowly relaxed in May 2024 for limited strikes near the Kharkiv front.[146] Ukraine's incursion into Russia's Kursk oblast in summer 2024 further complicated the decisionmaking process. Ultimately, in November 2024, Ukraine finally received long-sought approval from the United States and its European allies to use long-range missiles within Russia.[147] Because this decision still comes with limitations, Ukraine would need to rely on its own PGMs or drones for some long-range attacks within Russia, emphasizing the importance of sustaining native weapons production. Regardless of whether Ukraine is using U.S.-

---

[139] "Ukraine Launched a Massive Strike Using Rare Vilkha Rockets, Russia Claims," 2024.

[140] John Curtis and Claire Mills, *Military Assistance to Ukraine Since the Russian Invasion*, House of Commons Library, March 23, 2022.

[141] Christina L. Arabia, Andrew S. Bowen, and Cory Welt, *U.S. Security Assistance to Ukraine*, Congressional Research Service, IF12040, May 22, 2024.

[142] Terry Atlas, "U.S. Anti-Tank Missiles Headed to Ukraine," *Arms Control TODAY*, Vol. 48, April 2018.

[143] The United States' hesitancy has largely been motivated by concerns over escalating the conflict and provoking Russia into direct attacks on the North Atlantic Treaty Organization. See Isabelle Khurshudyan, Karen DeYoung, Alex Horton, and Karoun Demirjian, "Ukraine Wants More 'Game-Changer' HIMARS. The U.S. Says It's Complicated." *Washington Post*, July 24, 2022.

[144] Khurshudyan, DeYoung, et al., 2022.

[145] Stefan Korshak, "New Gliding Bombs and Precision-Guided Missiles Allow Ukraine to Reach Out and Hit Harder," *Kyiv Post*, March 7, 2023.

[146] Mary Bruce and Michelle Stoddart, "Biden Gave Ukraine Permission to Strike Inside Russia with US Weapons: Official," ABC News, May 30, 2024.

[147] Helene Cooper, "Ukraine Fires British Long-Range Missiles Into Russia," *New York Times*, November 20, 2024.

provided munitions or domestically produced Vilkha-M, the accuracy of its PGM attacks depends on the reliability of GPS-provided PNT signals, a challenge discussed in the next section.

Table 1. Timeline of Key U.S.-Provided Fire Capabilities

Initial Commitment Date	System Name	Description	Additional Information
April 2022	M777 Howitzer	Towed 155mm artillery piece	
June 2022	High Mobility Artillery Rocket Systems (HIMARS)	Wheeled multiple rocket launcher	
June 2022	M31 GMLRS	Precision-guided artillery rockets	Launched by HIMARS
September 2022	M982 Excalibur	Extended-range GPS-aided artillery shell	Launched by M777 Howitzer
December 2022	JDAM-ER	Kit for retrofitting "dumb" bombs with gliding wings and GPS guidance	
November 2023	GBU-39 SDB	Precision-guided air-dropped small bomb	
February 2024	GLSDB	Ground-launched version of GBU-39	Launched by HIMARS
April 2024	ATACMS	Long-range, GPS-aided tactical ballistic missile system	Launched by HIMARS

SOURCES: Features information from Mark F. Cancian and Chris H. Park, "What Is in the Ukraine Aid Package, and What Does It Mean for the Future of the War?" Center for Strategic and International Studies, May 1, 2024; Benton Coblentz, "US Expected to Decide Soon on Long-Range Missiles for Ukraine," Atlantic Council, September 12, 2023; C. Todd Lopez, "HIMARS, Excalibur Rounds Headed for Ukraine in $625 Million Security Assistance Package," U.S. Department of Defense, October 4, 2022; DoD, "$600 Million in Additional Security Assistance for Ukraine," September 15, 2022b; John Ismay, "What Are JDAMs? And What Will They Do for Ukraine?" *New York Times*, December 23, 2022b; James Bickerton, "U.S. Giving Ukraine Smart Bombs to Double Strike Range Against Russia," *Newsweek*, February 3, 2023; Isabelle Khurshudyan and Alex Horton, "Russia Jamming Leaves Some High-Tech U.S. Weapons Ineffective in Ukraine," *Washington Post*, May 24, 2024; Courtney Kube, "Ukraine Uses Long-Range ATACMS Against Russia for the First Time," NBC News, April 24, 2024.

Positioning, Navigation, and Timing Interference

Like communications satellites, GNSS spacecraft broadcast PNT signals using the electromagnetic spectrum. However, as these signals reach the earth's surface, they can be drowned out—or jammed—by terrestrial devices broadcasting "louder" interference signals at the same frequencies, making it difficult or impossible for nearby GNSS receivers to isolate the original PNT signal.[148] Jamming, a core tactic of electronic warfare (EW), is a particularly acute problem for PNT because GNSS satellites generally transmit weaker signals and over lower frequency bands compared with SATCOM systems.[149] This means that jamming GNSS services requires significantly less power

[148] Clayton Swope, Kari A. Bingen, Makena Young, Madeleine Chang, Stephanie Songer, and Jeremy Tammelleo, *Space Threat Assessment 2024*, Center for Strategic and International Studies, April 2024.

[149] "When GPS Fails, How Can Weapons Find Their Targets?" *The Economist*, July 14, 2023.

and can be achieved with relatively small, inexpensive, and easily deployed devices. Unlike jammers, which overwhelm GNSS signals, spoofing EW capabilities can "trick" GNSS receivers by injecting false PNT signals.[150] In the war in Ukraine, both Russia and Ukraine have used EW to disrupt all kinds of electromagnetic transmissions at different frequencies, including radio communications between soldiers, control linkages between pilots and drones, and tracking signals from air defense radars.[151] Additionally, both Russia and Ukraine have also successfully used spoofing to confuse and disable adversary GNSS-enabled capabilities.[152]

For years, Russia has been actively invested in advancing its EW capabilities, which continue to be employed on the battlefield in Ukraine.[153] From the first days of the invasion, recognizing the immense threat that Russia's EW capabilities pose to a wide variety of defensive operations, Ukraine resorted to a low-tech yet effective strategy of prioritizing the targeting and destroying of known Russian EW assets rather than focusing solely on developing counter–EW measures. In early March 2022, Ukraine's then–Minister of Defence Oleksii Reznikov urged Ukrainians to report these systems to the military, which would then track and attack them.[154] According to Ukraine's open-source intelligence group Molfar, from February 2022 through April 2023, Ukraine succeeded in destroying, damaging, or capturing 133 Russian EW systems or radars.[155]

The influx of U.S. PGMs initially provided a significant morale boost and enabled Ukrainian strikes that were more frequent, longer-range, and more precise. However, over time, their impact has reportedly been increasingly blunted by Russian GPS jamming.[156] As early as May 2023, reports surfaced that Russia had successfully extended its EW operations to disrupt GPS guidance on U.S.-provided munitions.[157] The same capabilities that Ukraine once urgently sought and heralded as "game-changers" were suddenly losing their tactical edge by becoming much less accurate.

---

[150] "Pokrova EW System Is a Real Game-Changer in Ukrainian Fight Against Shahed-136 Drones and Cruise Missiles, That Renders GPS Receivers Useless," Defense Express, November 4, 2023.

[151] Paul Mozur and Aaron Krolik, "The Invisible War in Ukraine Being Fought Over Radio Waves," *New York Times*, November 19, 2023; Hunter Stoll, John Hoehn, and William Courtney, "Air Defense Shapes Warfighting in Ukraine," RealClearDefense, February 22, 2024.

[152] David Hambling, "Ukraine Is Spoofing Russian Drones Out of the Sky," *Forbes*, April 21, 2023; Sam Skove, "Another US Precision-Guided Weapon Falls Prey to Russian Electronic Warfare, US Says," Defense One, April 28, 2024.

[153] Examples of research on this subject include Clint Reach, Alyssa Demus, Michelle Grisé, Khrystyna Holynska, Christopher Lynch, Dara Massicot, and David Woodworth, *Russia's Evolution Toward a Unified Strategic Operation: The Influence of Geography and Conventional Capacity*, RAND Corporation, RR-A1233-8, 2023; Sergey Sukhankin, "Blind, Confuse and Demoralize: Russian Electronic Warfare Operations in Donbas," Jamestown Foundation, August 27, 2021; Timothy Thomas, *Russia's Electronic Warfare Force: Blending Concepts with Capabilities*, MITRE Center for Technology and National Security, September 2020.

[154] "The Minister of Defense Urged Ukrainians to Destroy Russian Electronic Warfare and Reconnaissance Systems" ["Міністр оборони закликав українців знищувати російські системи радіоелектронної боротьби і розвідки"], Radio Liberty, March 9, 2022.

[155] Molfar, "Destruction of Electronic Warfare Equipment as a Prelude to a Counteroffensive. Analytics Confirms Dates of Counteroffensive Announced by the Ministry of Defense of Ukraine," undated.

[156] Mykhaylo Zabrodskyi, Jack Watling, Oleksandr V. Danylyuk, and Nick Reynolds, *Preliminary Lessons in Conventional Warfighting from Russia's Invasion of Ukraine: February–July 2022*, Royal United Services Institute for Defence and Security Studies, November 30, 2022.

[157] Alex Marquardt, Natasha Bertrand, and Zachary Cohen, "Russia's Jamming of US-Provided Rocket Systems Complicates Ukraine's War Effort," CNN, May 6, 2023.

Throughout the war, this has resulted in a continuous game of cat and mouse. Ukraine, the United States, and relevant systems manufacturers have worked to find countermeasures to protect PGMs from GPS interference. This effort involved modifying HIMARS rocket launchers, issuing software patches, "teaching" JDAMs to detect and circumvent jammers, and changing tactics to use U.S. PGMs equipped with GPS receivers that have proven more jam-resistant.[158] However, Russia has developed its own work-arounds to outfox Ukrainian and U.S. countermeasures.[159] For example, a Ukrainian military report found that the confirmed hit rate of GPS-guided Excalibur shells dropped from 55 percent in January 2023 to 6 percent in August 2023.[160] HIMARS-launched GMLRS and air-dropped JDAM bombs saw their own stark reductions in precision and effectiveness.[161]

As Ukraine began losing its PGM advantage because of Russian GPS interference, it once again resorted to asymmetric tactics as a possible countermeasure. This involved targeting Russian EW systems with drones immediately before launching GPS-guided missiles from HIMARS.[162] After some successes, Russia responded by increasingly deploying smaller and more-mobile EW systems that were harder to track and destroy.[163] However, Russia is not alone in its employment of EW—Ukraine has also been fielding its own EW capabilities to interfere with Russian attacks. These include such systems such Anklav and Bukovel-AD, which were developed by the Ukrainian company Proximus in 2014.[164] Capable of jamming GNSS signals, Bukovel-AD managed to down Russia's most-advanced drones, such as ZALA and Lancet; Russia acknowledged these events.[165] Bukovel-AD was also deployed near Ukraine's border when Russia started amassing its troops in 2021.[166] Moreover, in his widely-read November 2023 article in *The Economist*, then–Commander-in-Chief of

---

[158] For instance, Ukraine has expanded its usage of air-dropped GBU-39 bombs and curbed its usage of its ground-launched variant, GLSDB, which has proven more susceptible to interference (Khurshudyan and Horton, 2024). See also Sam Skove, "Russia Is Jamming US Precision Weapons in Ukraine, US General Says," Defense One, December 12, 2023b; Carlotta Gall and Vladyslav Golovin, "Some U.S. Weapons Stymied by Russian Jamming in Ukraine," *New York Times*, May 25, 2024; Thomas Withington, "Jamming JDAM: The Threat to US Munitions from Russian Electronic Warfare," Royal United Services Institute for Defence and Security Studies, June 6, 2023.

[159] Marquardt, Bertrand, and Cohen, 2023.

[160] Gall and Golovin, 2024.

[161] Abdujalil Abdurasulov, "Ukraine's Invisible Battle to Jam Russian Weapons," BBC News, August 3, 2023; Tom Porter, "Russia Is Jamming 'Sophisticated' US Weapons Being Used in Ukraine, Making Them Useless, Report Says," Business Insider, August 4, 2023; Skove, 2023b.

[162] Khurshudyan and Horton, 2024.

[163] Abdurasulov, 2023; Tom Porter, 2023.

[164] Mikhail Zhirohov, "Ukraine Grows EW Capability to Address Russia in Donbass," Janes, June 12, 2020; David Axe, "'Massive' Radio-Jamming Has Turned Krynky into a Robotic Kill-Zone for Hapless Russians. 'We Are Hampered.'" *Forbes*, January 5, 2024.

[165] "Ukrainian Bukovel-AD EW System 'Landed' Russian ZALA 421-16E2 UAV," Militarnyi, March 16, 2023; "The Defeat of the 'Lancet' of the EW Station 'Bukovel-AD' of the Armed Forces of Ukraine Got into the Frame," Military Review, June 27, 2023; Sunil J. B. Babu and Huw Williams, "Ukraine Conflict: Ukraine's Electronic Warfare Systems in Focus," Janes, March 4, 2022.

[166] "Ukrainian Bukovel-AD R4 EW-Systems Deployed near the Border," InformNapalm, November 21, 2021.

the Armed Forces of Ukraine Valery Zaluzhny discussed the nationwide deployment of Pokrova, an EW capability that spoofs GNSS-enabled systems.[167]

In mid-January 2024, it was also reported that Ukraine was able use EW to counter not only Russian drones but Russian-guided missiles as well. If this capability proves reliable and is consistently applied, some predict that it might become an inflection point for Ukraine.[168] Other forecasts remain less optimistic.[169] But just as Ukraine developed countermeasures for Russian EW tactics, Russia has also developed countermeasures against Ukrainian EW tactics. Reports have included Russian troops taking such makeshift measures as taping a Ukrainian SIM card and 4G modem to a drone's fuselage, thereby minimizing its dependence on satellite navigation.[170] As the battlefield becomes further saturated with EW, Russia and Ukraine continue to experiment with new solutions that avoid reliance on space-based radio signals that are vulnerable to EW. One such innovation that has been employed by both sides since at least summer 2024 is the fiber optic drone, which is thought to be jam resistant.[171]

Key Takeaways

As with SATCOM, Russia's longstanding underinvestment in GNSS and lagging PGM production capacity hamstrung its ability to leverage GLONASS-guided missiles early on in its war on Ukraine. Issues with GLONASS itself—as well as underdeveloped and undersupplied missile technology—constrained the reliability and availability of Russian guided munitions. However, Russia's more-recent successes with guided glide bombs, Shahed-type drones, and the launch of Zircon could signal a potential surge in its precision strike capabilities.

Russia also has benefited from its denial of Ukraine's PNT through EW, particularly by jamming the GPS signals needed to improve the precision of U.S.-provided munitions. Although Ukraine, the United States, and weapons manufacturers have developed some countermeasures against Russian interference, Ukraine continues to face serious accuracy challenges with many GPS-guided munitions. In addition to these existing frustrations, Ukraine has also experienced U.S. restrictions against using provided PGMs for long-range strikes in Russian territory, delays in aid deliveries, and uncertainty regarding future military assistance. This has compelled Ukraine to focus on enhancing its limited indigenous PGM production capacity and seek innovative solutions, such as jam-resistance drones.

---

[167] Valery Zaluzhny, "The Commander-in-Chief of Ukraine's Armed Forces on How to Win the War," *The Economist*, November 1, 2023; "Pokrova EW System Is a Real Game-Changer in Ukrainian Fight Against Shahed-136 Drones and Cruise Missiles, That Renders GPS Receivers Useless," 2023.

[168] Kateryna Stepanenko, Karolina Hird, Riley Bailey, Angelica Evans, and Frederick W. Kagan, *Russian Offensive Campaign Assessment, January 13, 2024*, Institute for the Study of War, January 13, 2024.

[169] Roman Olearchyk, "Military Briefing: Russia Has the Upper Hand in Electronic Warfare with Ukraine," *Financial Times*, January 6, 2024.

[170] United Kingdom Ministry of Defence GB [@DefenceHQ], "Latest Defence Intelligence update on the situation in Ukraine – 6 December 2023. Find out more about Defence Intelligence's use of language: ow.ly/Il6C50QfRHV UA #StandWithUkraine UA," post on the X platform, December 6, 2023.

[171] David Hambling, "Ukraine Fields Unjammable Fiber Optic FPV Attack Drone," *Forbes*, November 7, 2024.

Whether from observing the war in Ukraine or from other factors, U.S. officials have noted an "over-reliance" on GPS and greater attention to PNT. Russian activity in this domain points to greater challenges for the United States in future decades.[172]

Intelligence, Surveillance, and Reconnaissance

ISR describes the collection, analysis, and dissemination of diverse intelligence data across political, military, economic, and other domains to equip decisionmakers and planners with timely and accurate information.[173] Space-based remote sensing is especially important for contemporary militaries, given its ability to provide unparalleled coverage, persistence, and resolution. These satellites can enable continuous overhead monitoring of vast territories, detection of electronic signals, and precise geospatial mapping—making them pivotal for real-time situational awareness and longer-term strategic intelligence.

In this section, we explore the significant role space-based ISR has played throughout the war in Ukraine. We begin with a discussion of Russia and Ukraine's respective capabilities for collecting and analyzing satellite ISR products. Next, we examine the benefits and challenges of satellite intelligence sharing arrangements between the United States and Ukraine. We then discuss how commercial remote-sensing companies have shaped the war.

Russia's Space-Based Intelligence, Surveillance, and Reconnaissance

Much of Russia's space-based ISR capability depends on outdated technologies that significantly lag behind Western systems, including decades-old, Soviet-era designs. In the post-Soviet period, Russia largely focused on updates to the Soviet Union's flagship Yantar reconnaissance satellite series. This continued until as late as 2015, when Russia launched its final Yantar—Kosmos-2505, a Yantar-4K2M—which still remains operational.[174] Similar to all satellites in the Yantar series, the Yantar-4K2M uses physical film photography rather than the digital imaging and transmission technologies used by most ISR satellites.[175] This means that intelligence data from Yantar satellites could only be physically transmitted to earth through its film capsule, which is typically only collected after the entire roll is used.[176] This severely limits Yantar's value for rapid decisionmaking or real-time targeting. Furthermore, satellites in the Yantar line are not very durable and have relatively short lifespans.[177]

Around the same time as Yantar's final launch, Russia began finding success in deploying more-modern reconnaissance satellites. This includes the Persona series, a digital ISR platform capable of

[172] Dana A. Goward, "US Leaders Have Been Warned to Focus on GPS and PNT to Protect the Nation," *SpaceNews*, August 28, 2024.

[173] Joint Publication 3-14, 2023.

[174] "Russia Launches First Soyuz Rocket Since Space Station Mission Failure," NBC News, June 5, 2015.

[175] Anatoly Zak, "Kobalt-M Reconnaissance Satellite Series," webpage, RussianSpaceWeb.com, September 11, 2023.

[176] Zak, 2023.

[177] Burlaka, 2023.

collecting remote sensing data and transmitting it in real time to Russian ground stations for analysis. Although the first Persona satellite was launched in 2008, it failed to return useful imagery because of a technical malfunction.[178] Then, following a series of delays, two Persona satellites were eventually successfully deployed—one in 2013 and one in 2015—and remain the only Persona satellites known to have been launched.[179] Furthermore, Persona satellites, which use optical sensors, are highly sensitive to weather and daylight conditions and are unable to "see" through cloud cover or at nighttime.[180] Many Western satellites—both government and commercial—employ synthetic aperture radar (SAR) technology to overcome these shortcomings. Russia has one confirmed SAR-capable reconnaissance satellite, the Kondor-E, but it was only launched as an experiment; it entered orbit in 2014 and is now substantially past its expected lifespan of five years.[181]

In fact, most of Russia's existing ISR satellites are nearing the end of or have already exceeded their operational lifespans, so it remains unclear how heavily Russia has been able to rely on space-based ISR during the war in Ukraine. However, Russia continued its efforts to enhance its optical imaging capabilities, which included launching two additional Bars-M satellites since the full-scale invasion.[182]

Ukraine's Space-Based Intelligence, Surveillance, and Reconnaissance

Although Ukraine had ambitions of developing fully indigenous space-based ISR capabilities—including its own launch and satellite programs—roadblocks and delays have largely narrowed its focus on the analysis of externally provided remote sensing data. This effort has involved the establishment of the National Center Space Facilities Control and Test Center (NSFCTC), an agency operating under the auspices of the State Space Agency of Ukraine. Although the NSFCTC is part of a nonmilitary institution and includes a civilian component in its portfolio, it was primarily tasked with advancing Ukraine's defense-related ISR capabilities.[183] However, prior to 2014, Ukraine's ISR analysis capabilities were mainly used for civilian purposes, such as agriculture management or forest fire monitoring.[184]

Following Russia's initial invasion in 2014, defense applications became a primary focus of Ukraine's ISR efforts; agreements were struck with both commercial and government remote sensing providers for the procurement of raw satellite data. For instance, in 2019, the State Space Agency of

---

[178] Natalia Kozlova, "Suppliers Are to Blame for the Loss of the Military Satellite 'Persona'" ["В гибели военного спутника 'Персона' виноваты поставщики"], *Russian Gazette* [*Российская газета*], February 11, 2009.

[179] "Persona Satellite Overview," Spaceflight101.com, undated.

[180] Rod Thornton and Marina Miron, *Russian Military Space-Related Capabilities: The Vital Deterrence Role of Counterspace Weapons (Part I)*, Freeman Air and Space Institute, May 2024; Connell, 2023.

[181] "Dual-Purpose Radar Satellite 'Kondor-E' Deployed to Target Orbit" ["Радиолокационный спутник двойного назначения 'Кондор-Э' выведен на целевую орбиту"], *Vedomosti*, December 19, 2014.

[182] Paweł Bernat, "The Evolution of Russian ISR Satellite Capabilities Following the Full-Scale Aggression on Ukraine (2022–2024)," paper presented at the Combating Threats from the Air: Current Issues and Ways to Solve Them conference, Kyiv, Ukraine, October 2024.

[183] National Space Management and Testing Center, "Structure of NSFCTC," webpage, undated.

[184] Cabinet of Ministers of Ukraine, *Public Report of the Head of the State Space Agency of Ukraine for 2019* [*Публічний звіт Голови Державного космічного агентства України за 2019 рік*], February 24, 2020.

Ukraine signed a contract with the European Union Commission to access data collected by Sentinel satellites as part of the Copernicus program.[185] Further expanding its capabilities, NSFCTC entered into a contract in August 2021 with the South African company Dragonfly Aerospace to obtain their remote sensing data.[186] Ukraine then analyzed and used this data for various purposes, including national security.

In the intervening years between Russia's initial invasion in 2014 and the full-scale invasion in 2022, access to satellite data proved to be exceedingly valuable, although its absence highlighted significant informational gaps.[187] In 2017, several explosions occurred at Ukraine's arms depots in the eastern and central regions of the country. Ukraine attributed these incidents to Russia's drones but was unable to provide conclusive evidence, as none of the Western surveillance satellites were well-positioned to detect them.[188] Ukraine claimed that the trajectory and location of surveillance satellites over those areas suggested Russia's involvement because it managed to guide its drones undetected.[189]

Similarly, Ukraine debunked Russia's allegations that Ukrainian fighters were responsible for downing the Dutch passenger aircraft MH17 by showing that Russian satellites were not positioned over the area at the time of the incident; therefore, Russia could not have known whether Ukrainian fighters were responsible.[190] On several occasions, the Ukrainian military openly expressed concerns about Russian military personnel and equipment amassing at its borders. Using remote sensing data, Ukraine closely monitored Russia's militarization of the Crimean Peninsula, locating and regularly reporting on newly constructed military bases, airfields, and storage facilities.[191] In late 2018, Russia was observed orchestrating its largest buildup since the Crimea occupation in 2014. Ukrainian military leadership was able to monitor the situation through acquired satellite imagery and develop scenarios for possible responses.[192]

In the lead-up to Russia's full-scale invasion of Ukraine, however, it was U.S. intelligence products—and images produced by U.S. space companies—that played a pivotal role in tracking Russian movements. For weeks before the invasion, Washington publicly revealed unprecedented amounts of detailed intelligence warning of the impending invasion. Indeed, in January 2022, U.S. President Joe Biden bluntly warned that Russia "'will move in' to Ukraine."[193] Intelligence disclosures

---

[185] Svitlana Hudkova, "One Cannot Do Without Elon Musk: With Whom and How Ukraine Will Launch Satellites" ["Без Ілона Маска не обійтися: з ким і як Україна буде запускати супутники"], Apostrophe [Апостроф], May 31, 2021; European Space Agency, "Southern Ukraine," webpage, May 1, 2020.

[186] "Dragonfly Aerospace Provided Services to Ukraine in Remote Sensing of the Earth" ["Dragonfly Aerospace надала послуги Україні у дистанційному зондуванні Землі]," Militarnyi, September 19, 2021.

[187] Ihor Tokar, "Declassified. How Does Ukraine Monitor Russia in Crimea from Space?" ["Розсекретили. Як Україна стежить за Росією в Криму з космосу?"] Radio Liberty, February 7, 2021.

[188] "Lutsenko on TSN. Tyzhden Showed Evidence of Sabotage at the Warehouses in Balakliia" ["Луценко в ефірі ТСН. Тижня показав докази диверсії на складах у Балаклії"], Television News Service, October 2, 2017.

[189] "Ukraine's Military Space. Part 2" ["Військовий космос України. Частина 2"], Militarnyi, December 15, 2017.

[190] "Ukraine's Military Space. Part 2," 2017.

[191] "Ukrainian Space Center Showed How Russia Is Increasing Its Military Presence in Crimea" ["В українському космічному центрі показали, як Росія нарощує військову присутність у Криму"], Radio Liberty, February 8, 2021.

[192] Dylan Malyasov, "Satellite Imagery Shows Hundreds of Russian Tanks near the Border with Ukraine," Defence Blog, August 30, 2020.

[193] Kevin Liptak, "Biden Predicts Russia 'Will Move in' to Ukraine, but Says 'Minor Incursion' May Prompt Discussion over Consequences," CNN, January 19, 2022.

included information on Russian troop movements and the Kremlin's plans for false flag attacks.[194] The Western commercial companies Maxar and Planet Labs provided images of Russian buildups along the Ukrainian border to such major media outlets as the *New York Times* and *Wall Street Journal*.[195] These companies' images were also privately purchased by the U.S. National Geospatial-Intelligence Agency (NGA) and National Reconnaissance Office; according to an NGA official, the images were "able to flow directly to those who need it, EUCOM [U.S. European Command], NATO and directly to Ukrainians."[196] Although the assessments concluded that Russia had already decided to invade and did not dissuade Russian actions, U.S. intelligence sharing might have reduced the effectiveness of Russia's attack by prompting Ukraine to make preparations.[197] Additionally, U.S. intelligence disclosures helped facilitate consensus building among allies, such as the United Kingdom and the Baltic states.[198]

As Russia began advancing—and thereafter throughout the conflict—the United States continued to share its intelligence with Ukraine. In the earliest days of the conflict, then–White House Press Secretary Jen Psaki confirmed that the U.S. "[had] consistently been sharing intelligence that includes information the Ukrainians can use to inform and develop their military response to Russia's invasion."[199] However, around this same time, several members of Congress began publicly expressing concerns that the intelligence provided was too limited for Ukraine's evolving warfighting objectives. This prompted the White House to reportedly revise and expand its intelligence sharing approach for Ukraine.[200] This included an April 2022 guidance clarification that explicitly empowered U.S. intelligence bodies to equip Ukraine with intelligence for operations in Russian-occupied Donbas and Crimea—a change confirmed by U.S. Secretary of Defense Lloyd Austin in his testimony to the U.S. Senate Committee on Armed Services.[201]

Moreover, on signing a bilateral security agreement with Ukraine in June 2024, President Biden committed to "expanding intelligence-sharing" even further.[202] Shared U.S. intelligence has explicitly included space-based ISR products; for example, DoD has confirmed its provision of "[s]atellite

---

[194] Amy Zegart, "Open Secrets: Ukraine and the Next Intelligence Revolution," *Foreign Affairs*, Vol. 102, No. 1, January/February 2023.

[195] Michael Schwirtz, Scott Reinhard, and Josh Holder, "How Russia Has Increased Its Military Buildup," *New York Times*, January 27, 2022; Warren P. Strobel and Robert Wall, "Ukraine War Puts Spy Satellites for Hire in the Spotlight," *Wall Street Journal*, May 1, 2022.

[196] David Gauthier, as quoted in Sandra Erwin, "As Russia Prepared to Invade, U.S. Opened Commercial Imagery Pipeline to Ukraine," *SpaceNews*, April 6, 2022.

[197] Michael Schwirtz, Anton Troianovski, Yousur Al-Hlou, Masha Froliak, Adam Entous, and Thomas Gibbons-Neff, "Putin's War," *New York Times*, December 16, 2022; Egle E. Murauskaite, *U.S. Assistance to Ukraine in the Information Space: Intelligence, Cyber, and Signaling*, Asymmetric Threats Analysis Center, University of Maryland, February 2023.

[198] Murauskaite, 2023.

[199] Idrees Ali and Phil Stewart, "U.S. Providing Intelligence to Ukraine, Officials Say," Reuters, March 3, 2022.

[200] Ken Dilanian, "Biden Administration Walks Fine Line on Intelligence-Sharing with Ukraine," NBC News, March 4, 2022.

[201] Michael R. Gordon, Warren P. Strobel, and Vivian Salama, "Biden Administration to Provide Ukraine with More Intelligence, Heavier Weapons to Fight Russia," *Wall Street Journal*, April 13, 2022; Lloyd Austin III, testimony for the hearing on the Department of Defense Budget Posture in Review of the Defense Authorization Request for Fiscal Year 2023 and the Future Years Defense Program, U.S. Senate Committee on Armed Services, April 7, 2022.

[202] Joseph Biden and Volodymyr Zelenskyy, transcript of remarks given at joint press conference in Fasano, Italy, White House, June 13, 2024.

imagery and analysis capability" and direct "[c]ommercial satellite imagery services" in Ukrainian assistance packages.[203] U.S. intelligence on key aspects of Russian military movements and supplies has reportedly played an integral role in Ukrainian planning of its counteroffensive strategies, including targeting and killing Russian generals.[204]

Commercial Involvement

In addition to satellite intelligence provided by the U.S. government, Ukraine has also secured access to commercial ISR data directly from U.S. space companies. Four U.S.-based remote sensing companies—Maxar, Planet Labs, Capella, and BlackSky—proved particularly critical for Ukraine's war efforts, with some even adjusting their operations to better assist Ukraine.[205] For example, days before the full-scale invasion, BlackSky changed the planned orbit of two imaging satellites—originally set to launch weeks later—to maximize their coverage over Ukraine. The company was then able to provide Ukraine with imagery within 24 hours of the April launch.[206]

HawkEye 360, another U.S. geospatial analytics company, has filled a different niche for Ukraine. By using satellite sensors that detect radio frequency signals—a distinct phenomenology from other companies' optical imaging satellites—HawkEye spacecraft can identify and geolocate activities, such as GPS interference. This data has enabled Ukraine to track and target Russian forces by following their associated jamming signals.[207]

Fearing interruptions in the provision of the imagery by Western governments or commercial companies at critical moments, Ukraine has also sought to secure more long-term partnerships. This was facilitated by Ukrainian civil society, which had already been actively involved in providing equipment to Ukraine's military. In August 2022, the Serhiy Prytula Charity Foundation, a Ukrainian charity, orchestrated a highly publicized crowdfunding effort for Ukraine to purchase Bayraktar TB2 drones.[208] However, when the Turkish manufacturer Baykar instead offered to donate the systems, Ukraine redirected the collected funds to secure continuous access to satellite imagery from the

---

[203] DoD, "Fact Sheet on U.S. Security Assistance for Ukraine (Roll Up) as of April 21, 2022," press release, April 22, 2022a; DoD, "Biden Administration Announces Additional Security Assistance for Ukraine," press release, May 9, 2023.

[204] Julian E. Barnes, Helene Cooper, and Eric Schmitt, "U.S. Intelligence Is Helping Ukraine Kill Russian Generals, Officials Say," *New York Times*, May 4, 2022; Julian E. Barnes and Helene Cooper, "Ukrainian Officials Drew on U.S. Intelligence to Plan Counteroffensive," *New York Times*, September 10, 2022.

[205] "What Pentagon Leaders Say They Have Learned One Year on from the Battle in Ukraine," Breaking Defense, February 23, 2023.

[206] Strobel and Wall, 2022.

[207] Strobel and Wall, 2022.

[208] "What Is the ICEYE Satellite Purchased with Bayraktar TB2 Funds and Why It Really Makes a Difference," Defense Express, August 19, 2022.

Finnish remote sensing company ICEYE.[209] ICEYE's constellation is importantly SAR-capable, providing Ukraine with high-quality imagery at any time and through any weather.[210]

Within the first months of using ICEYE's data, the Ukraine's Ministry of Defence claimed that Ukraine was able to identify the location of over 7,000 Russian military equipment sites and troop positions.[211] Ukraine's military was then able to conduct and confirm the destruction of hundreds of Russian assets, including fighters and advanced missile launchers.[212] ICEYE and other commercial providers have also enhanced Ukraine's drone missions by providing a better understanding of operational environments.[213] In July 2024, ICEYE and Ukraine's Ministry of Defence signed a memorandum to further strengthen their cooperation in remote sensing and to work on preventing Russia's access to imagery that could enhance its targeting of Ukraine's territory.[214]

Commercial data provides a useful complement to other intelligence sources for Ukraine. By one account, the greatest advantage gained through commercial remote sensing has been the ability to monitor the situation within Russia because Western partners had largely supplied only imagery of Ukraine. Data from such companies as ICEYE became instrumental for Ukraine to observe the effects of strikes on targets inside Russian territory.[215] Overhead intelligence on battlefield conditions, in occupied territories, and in de-occupied territories also proved critical for Ukraine's information campaigns. For example, images procured by Planet Labs of burning helicopters in Chornobaivka helped Ukraine debunk Russia's flawed claims about its early offensives.[216] Planet Labs' data further assisted Ukraine in documenting the destruction of an expensive and supposedly advanced Russian missile defense S-400 system in Crimea in August 2023.[217] Also in Crimea, Ukraine was able to show that Russia had constructed extensive defensive lines, which contradicted Russian assertions that these territories were entirely secure and beyond reach.[218]

---

[209] Baykar, "Fundraising Campaigns for Purchase of Türkiye's Bayraktar TB2 Drone Spreading in West," July 23, 2022; ICEYE, "ICEYE Signs Contract to Provide Government of Ukraine with Access to Its SAR Satellite Constellation," press release, August 18, 2022.

[210] "Prytula Provided the Armed Forces of Ukraine Access to the ICEYE Satellite Imaging Database" ["Притула передав ЗСУ доступ до бази даних супутникових знімків ICEYE"], Detector Media, August 18, 2022.

[211] "How Many Objects of the Occupiers Did the 'People's Satellite' ICEYE Help Discover" ["Скільки об'єктів окупантів допоміг виявити 'народний супутник' ICEYE"], Texty.org.ua, March 9, 2023.

[212] Valeriia Bunyak, "GUR: ICEYE Satellite, Purchased with Donations from Ukrainians, Helped Destroy Thousands of Enemy Units" ["ГУР: Супутник ICEYE, придбаний за пожертви українців, допоміг знищити тисячі одиниць ворожої техніки"], Detector Media, March 9, 2023.

[213] Mykhailo Glukhovsky, "Lieutenant Colonel Honchar: Data from 'Pritula's Satellite' Monopolized by One Entity" ["Підполковник Гончар: Дані з 'супутника Притули' монополізувала одна структура"], *Glavkom*, September 13, 2022.

[214] "ICEYE and the Ministry of Defence of Ukraine Sign a Memorandum of Cooperation," Defence Industry Europe, July 8, 2024.

[215] Halyna Tereshchuk, "Space Wars. Is Russia Capable of Using Nuclear Weapons in Space" ["Космічні війни. Чи здатна Росія застосувати ядерну зброю у космосі"], Radio Liberty, March 2, 2024.

[216] "Two Years of the Russian-Ukrainian War Through the Lens of Satellites," Universe Space Tech, February 23, 2024.

[217] "Satellite Image of the Aftermath of the Destruction of the S-400 Air Defense System in Crimea Appears" ["З'явився супутниковий знімок наслідків знищення ЗРК С-400 у Криму"], Militarnyi, August 28, 2023.

[218] Anzhelika Rudenko and Aleksina Dorohan, "'Why Trenches, Are We Expecting a War?' Defensive Fortifications in Crimea: What Can Be Seen from the Satellite?" ["'Навіщо окопи, чекаємо на війну?' Оборонні укріплення в Криму. Що видно із супутника?"], Crimea.Realities, April 5, 2023.

Commercial satellite imagery has also helped the West monitor other Russian activities during the war in Ukraine. For example, as discussions about the fragility of the Russian economy and military industrial base started to emerge, space-based ISR helped shed light on the measures Russia was taking to evade Western sanctions.[219] Commercial data was useful in exposing the routes and means by which Russia was exporting oil.[220] Despite Russia's repeated denials, Ukraine was able to document instances of Russia loading Ukrainian grain onto its ships in the Black Sea.[221] Multiple investigative reports used commercial remote sensing products to show North Korea transferring military systems to Russia.[222]

However, as with other commercial space services involved in the war, Russia has reportedly been able to circumvent U.S. restrictions and make use of space imagery from U.S. companies.[223] According to one Ukrainian military source, this intelligence has repeatedly informed Russia's targeting efforts, purportedly enabling hundreds of missile attacks within Ukraine. This concern has driven the Ukrainian government to attempt a crackdown on select commercially provided satellite images, including a memorandum of understanding between Ukraine's Ministry of Defence and multiple remote sensing companies to restrict the distribution of images containing Ukraine.[224] Nonetheless, given Russia's demonstrated knack for circumventing sanctions and other constraints—particularly in the space domain—these efforts could prove quite challenging, and Russia could continue using commercial vendors to compensate for its space-based ISR shortcomings.

Key Takeaways

Although Russia's dilapidated space-based ISR capabilities struggled to meaningfully support Russian warfighting, satellite intelligence—and especially commercial remote sensing products—has played an unprecedented role in shaping the conflict for Ukraine and the West. Advanced ISR satellites have helped track and anticipate Russian troop and material movements, improved situational awareness, informed targeting, built consensus among allies, and cultivated international support. Ukraine has sourced its satellite intelligence from governments (especially that of the United States) and several commercial vendors, leveraging its previous investments in remote sensing analysis capabilities. To cover gaps in intelligence received from government partners, Ukraine has turned to commercial remote sensing data from such providers as Maxar, Planet Labs, and ICEYE. The imagery from these companies not only alerted the world to Russia's pre-invasion buildup but has since

---

219 Benjamin Schmitt, "Catching Russia Sanctions Busters – From Orbit," Center for European Policy Analysis, June 6, 2023.

220 Christiaan Triebert, Blacki Migliozzi, Alexander Cardia, Muyi Xiao, and David Botti, "Fake Signals and American Insurance: How a Dark Fleet Moves Russian Oil," *New York Times*, May 30, 2023.

221 "Two Years of the Russian-Ukrainian War Through the Lens of Satellites," 2024.

222 Michelle Ye Hee Lee and Joyce Sohyun Lee, "North Korea May Be Sending Arms to Russia for Ukraine War, Images Suggest," *Washington Post*, October 16, 2023; Andrew Roth, "UK Sends UN Experts Photographs of North Korean Shipments to Russia," *The Guardian*, January 22, 2024.

223 Wood, 2024.

224 Dmytro Basmat, "Defense Ministry Working to Limit Satellite Imaging of Ukrainian Territory," *Kyiv Independent*, May 2, 2024.

assisted in documenting the destruction in Ukraine, thereby shaping the public narrative and catalyzing allied support. Importantly, the sheer number of satellite intelligence sources has helped obviate any "lock-in" issues with overreliance on a single provider, which has complicated Ukraine's experience in other such space mission areas as SATCOM. However, the multitude of providers also complicates efforts to restrict unauthorized use.

Conclusion

From Starlink's role in enabling Ukrainian warfighting to the cat-and-mouse game surrounding Ukraine's employment of GPS-guided munitions and the importance of commercial imagery in improving battlefield awareness, the availability and denial of space-based services have played a major role in the war in Ukraine. Particularly at the war's outset, Russia's underinvestment in space capabilities since the fall of the Soviet Union likely limited its ability to leverage space services to the same degree as Ukraine, which has effectively drawn on Western-provided or commercially available technologies. However, Russia has shown some progress in its space-enabled warfighting and has successfully disrupted Ukraine's access to space services periodically, including through cyberattacks and jamming.

The observations from this open-source review yield several findings and recommendations for the broader policy community about the role of space in future conflicts and ways in which the United States can draw lessons from the conflict to prepare for such eventualities.

Findings

Finding 1. Key Space Services Will Likely Be Disrupted in Future Conflicts

The war in Ukraine demonstrates the growing effect of the use or denial of space services on battlefield outcomes, as demonstrated by cyberattacks on Viasat, GPS jamming, and additional threats from Russia. Publicly available Defense Intelligence Agency reporting indicates that China possesses extensive space and counterspace capabilities, incorporates the jamming of space services into its exercises, and is probably developing jammers targeting SAR and SATCOM.[225] These observations suggest that the United States and its allies might face further disruption of space services in a potential future conflict with China.[226] The experience in Ukraine suggests that reversible attacks, such as cyberattacks and EW, might come relatively early in a crisis or conflict and reinforces the importance of operating space services that can withstand these or other attacks. A variety of DoD policies and investments have already prioritized multi-orbit and proliferated systems, and the experience in Ukraine reinforces those priorities.[227]

---

[225] Defense Intelligence Agency, *Challenges to Security in Space: Space Reliance in an Era of Competition and Expansion*, 2022, pp. 11–18.

[226] For a discussion of China's potential employment of space capabilities in crisis and conflict, see Howard Wang, Gregory Graff, and Alexis Dale-Huang, *China's Growing Risk Tolerance in Space: People's Liberation Army Perspectives and Escalation Dynamics*, RAND Corporation, RR-A2313-2, 2024.

[227] See, for example, USSF Gen Chance Saltzman's remarks on "competitive endurance" (Joseph Clark, "Space Force General Outlines U.S. Approach to Maintaining Space Superiority," U.S. Department of Defense, March 28, 2024).

Finding 2. Widespread Availability of Space Services Increases Transparency of the Battlefield Prior to and During Conflict

Commercial imagery and declassified intelligence helped to provide surprisingly timely warnings of Russia's attack. Once the war began, Ukraine quickly benefited from its ability to leverage commercial imagery in combination with intelligence shared by its Western partners. This is not to say that all actors have similar collection capabilities; Russia's national collection had degraded prior to the war, and some nations might have greater capabilities than others. However, the widespread availability of commercial imagery or of released national imagery will likely make future large-scale conflict relatively more visible in open sources.

Finding 3. Commercial Space Services Bring Value and Vulnerability

Relatedly, the war in Ukraine shows the value and vulnerability of commercially provided space services for countries embroiled in conflict. Ukraine's ability to leverage Western commercial services enhanced its warfighting and largely obviated the need to build native space capabilities. It also meant that Ukraine did not require national defense organizations dedicated to space—like those established by the United States—and instead organized its forces around the employment of EW, UAVs, and other capabilities. However, echoing our second finding, the widespread availability of commercial capabilities has also allowed Russia to find ways of leveraging these capabilities and partly overcome its own space shortcomings.

Another concern is that commercial providers are independent actors and might have different incentives from their national customers. Some of the specific challenges seen in Ukraine are likely unique to this conflict, such as the concerns around SpaceX unilaterally deciding where and when Starlink would be available to Ukrainian troops. This issue was largely made possible because of the voluntary and informal nature of SpaceX's initial commitment, subjecting its provision of Starlink to no formal conditions or direct government oversight. Space services would typically be provided under contract, and, indeed, SpaceX did eventually agree to contract with DoD. Going forward, DoD has stated its commitment to using "the full range of available financial, contractual, and policy tools to rapidly field and scale commercial technology . . . ."[228] Of course, this does not eliminate the possibility of any future disputes, but the particular challenges experienced with SpaceX early in the Ukraine war might be unique. However, Taiwan and other countries' hesitations around SpaceX highlights how reliability concerns make countries cautious of relying on a single commercial provider.

The war in Ukraine demonstrates how commercial space capabilities might be particularly vulnerable to attacks prior to the start of armed conflict because attacks on these systems might present lower escalation risks.[229] Russia's threats to attack commercial providers and its early cyberattack on Viasat are notable examples of how commercial vulnerabilities can affect warfighting. These events underscore the uncertainties of how DoD or other actors will respond to attacks on

---

[228] DoD, 2024, p. 4.

[229] Alexandra T. Evans, Andrew Radin, Katie Feistel, Krista Langeland, Bruce McClintock, and Howard Wang, *Space Strategic Stability: Assessing U.S. Concepts and Approaches*, RAND Corporation, RR-A2313-1, 2024.

commercial space systems; this question is raised—but not fully addressed—in both DoD's and the USSF's recent commercial space strategies.[230]

Finding 4. Future Warfare Will Include Opportunities and Challenges in Different Space Mission Areas

Turning to the lessons of specific mission areas, the war in Ukraine offers some important indications of future opportunities and challenges.

For SATCOM, the war has demonstrated the potential utility of proliferated SATCOM architectures for enabling terrestrial forces. Russia's apparent difficulty in jamming or otherwise denying Starlink services shows how governments can benefit from the added reliability of proliferated satellite architectures. However, the Viasat hack and lingering concerns over Starlink's reliability also highlight the importance of diversification, given the vulnerabilities inherent to relying on any single SATCOM provider.

Regarding PNT, the war in Ukraine has highlighted the critical warfighting capabilities enabled by space-provided GNSS signals, the significant challenges that these signals will continue to face from terrestrial EW operations, and the importance of partners generating their own capacity to produce guided munitions. Although Ukraine's native production of GPS-aided weapons has been limited, it has given Ukrainian commanders more flexibility to strike a wider variety of targets.

Lastly, ISR-specific lessons include the ability of commercial ISR products to substitute for national capabilities, at least for close U.S. partners. For many countries, the experience in Ukraine likely signals a reduction in the importance of fielding native space-based ISR capabilities if they can instead count on intelligence made available by Western companies and governments.

Across multiple mission areas, the availability of space services from outside organizations ostensibly reduces the need for countries, such as Ukraine, to organize explicit space capabilities. Therefore, countries can instead subordinate space forces underneath EW, intelligence, or other more-traditional capability areas. However, an overdependence on shared capabilities can still constrain U.S. partners' freedom of action. Finally, the rise of commercial imagery also creates risks that adversaries will exploit such imagery to the detriment of its intended user. Space capabilities, including ISR, will increasingly require careful attention to limit the risk of misuse or cyber penetration.

Recommendations

Given the constraints of this report in drawing only from open sources, the specific recommendations for DoD remain necessarily general given the limited information about specific systems and programs that are in the public domain. However, our findings still yield some concrete and relevant priority areas for the United States, its allies, and partner nations going forward.

---

230 USSF, 2024; DoD, 2024.

Recommendation 1. Plan for Leveraging Commercial Space Assets to Support Allies and Partners Leading up to and During Future Conflicts

The war in Ukraine provides useful experience for how to effectively leverage commercial resources to support partners. The U.S. government and commercial space providers should strategically prepare for delivering these services in support of U.S. allies and partners in the future. Wargames or other planning exercises can help decisionmakers think through what and when commercial services might be needed by different partners to ensure they are in place prior to the start of a crisis or conflict.

Recommendation 2. Continue Developing Robust Contract Arrangements with Commercial Space Providers and More Clearly Articulate U.S. Responses to Potential Threats to Commercial Space Systems

The United States, its allies, and its partners should ensure that contractual arrangements with commercial providers cover a wide variety of contingencies—including adversary kinetic or nonkinetic attacks—and should work with companies to anticipate how these contingencies might affect future service provision. Russian threats and attacks during the war in Ukraine can serve as motivation for such preparations. The absence of a formal contract arrangement with SpaceX created significant uncertainty, and DoD policymakers should take care to avoid such informal arrangements in the future. This priority is already recognized in U.S. commercial space strategy documents and initiatives, particularly in the DoD and USSF commercial space strategies; the Commercial Space Protection Tri-Seal Strategic Framework, which brings together NGA, the National Reconnaissance Office, and U.S. Space Command; and the creation of the Commercial Augmentation Space Reserve.[231] When appropriate, DoD and other U.S. government stakeholders can work closely with commercial and international partners to assuage concerns about the reliability of U.S. companies. Privately, if not publicly, discussing explicit U.S. responses to different forms of adversary attacks in advance could also help mitigate concerns. U.S. strategy documents identify potential options, such as "the use of military force to protect and defend commercial assets" and indemnification for commercial providers who are attacked.[232] However, policy statements do not necessarily translate into real-world actions, and DoD should do more to give commercial actors confidence in DoD protections and responses.

Recommendation 3. Mission Area Specific Recommendations

The dynamics of the war in Ukraine yield several recommendations for specific mission areas, although many of these efforts are already underway.

For SATCOM, DoD should continue its ongoing efforts to build resilient, multi-orbit, proliferated architectures. The war simultaneously points to the value of proliferated low earth orbit architectures and the importance of diversification. A variety of available communication pathways—

---

[231] DoD, 2024; USSF, 2024; Courtney Albon, "Space Force Plans to Award 20 Contracts for Commercial Reserve by 2026," Defense News, November 22, 2024; National Reconnaissance Office, "Tri-Seal Commercial Space Protection Framework," August 31, 2023.

[232] DoD, 2024, pp. 3–4.

including both military-owned systems and commercial services—would likely make it more difficult for an adversary to disrupt communications as seen in Ukraine.

For PNT, given likely adversary investments in GPS jamming or spoofing technologies, the United States, its allies, and its partners should innovate to identify alternative technologies to ensure that platforms and munitions remain effective. DoD has pursued greater GPS resilience through additional satellites and an additional military signal called M-code, but questions about receivers remain. A 2023 U.S. Government Accountability Office report noted past delays with addressing new GPS technology concerns around whether there was "a sound business case" for some planned receivers.[233] Alternative PNT options other than GPS can provide military users with additional options in the face of adversary disruption.[234]

For ISR, the United States should create clear expectations for the commercial or shared services that partners and allies can depend on in crisis or conflict. Then, those services can prepare accordingly while recognizing that each event is different and that actual provision of services might vary. Commercial providers should invest in basic cybersecurity protection to ensure that their systems are hardened against adversary attacks and to limit the risk of misuse of their services by adversaries.

---

[233] U.S. Government Accountability Office, *GPS Modernization: Space Force Should Reassess Requirements for Satellites and Handheld Devices*, GAO-23-106018, June 2023, p. 36.

[234] U.S. Government Accountability Office, *GPS Alternatives: DoD Is Developing Navigation Systems but Is Not Measuring Overall Progress*, GAO-22-106010, August 2022.

Abbreviations

DoD	U.S. Department of Defense
ESSS	Integrated Satellite Communication System
EW	electronic warfare
GLONASS	Global Navigation Satellite System
GMLRS	Guided Multiple Launch Rocket Systems
GNSS	global navigation satellite systems
GPS	Global Positioning System
HIMARS	High Mobility Artillery Rocket Systems
ISR	intelligence, surveillance, and reconnaissance
NGA	U.S. National Geospatial-Intelligence Agency
NSFCTC	National Center Space Facilities Control and Test Center
PGM	precision-guided munition
PNT	positioning, navigation, and timing
SAR	synthetic aperture radar
SATCOM	satellite communications
UAV	unmanned aerial vehicle
USSF	U.S. Space Force

Bibliography

Unless otherwise indicated, the authors of this report provided the translations of bibliographic details for the non-English sources included in this report. To support conventions for alphabetizing, sources in Russian and Ukrainian are introduced with and organized according to their English translations. The original rendering in Russian and Ukrainian appears in brackets after the English translation.

Abdurasulov, Abdujalil, "Ukraine's Invisible Battle to Jam Russian Weapons," BBC News, August 3, 2023.

Advanced Navigation, "Global Navigation Satellite System (GNSS) and Satellite Navigation Explained," March 8, 2023.

Albon, Courtney, "Space Force Plans to Award 20 Contracts for Commercial Reserve by 2026," Defense News, November 22, 2024.

Ali, Idrees, and Phil Stewart, "U.S. Providing Intelligence to Ukraine, Officials Say," Reuters, March 3, 2022.

Antoniuk, Daryna, "How Elon Musk's Starlink Satellite Internet Keeps Ukraine Online," *Kyiv Independent*, September 3, 2022.

Arabia, Christina L., Andrew S. Bowen, and Cory Welt, *U.S. Security Assistance to Ukraine*, Congressional Research Service, IF12040, updated May 22, 2024.

Atlas, Terry, "U.S. Anti-Tank Missiles Headed to Ukraine," *Arms Control TODAY*, Vol. 48, April 2018.

Austin, Lloyd, III, testimony for the hearing on the Department of Defense Budget Posture in Review of the Defense Authorization Request for Fiscal Year 2023 and the Future Years Defense Program, U.S. Senate Committee on Armed Services, April 7, 2022.

Axe, David, "Ukraine's GPS-Guided Bombs Work Even When You Jam Their GPS," *Forbes*, November 1, 2023.

Axe, David, "'Massive' Radio-Jamming Has Turned Krynky into a Robotic Kill-Zone for Hapless Russians. 'We Are Hampered.'" *Forbes*, January 5, 2024.

B., Volodymyr, "Vilkha: Ghost Weapon or Game Changer?" ["Вільха: зброя-привид чи gamechanger?"], Militarnyi, October 23, 2023.

Babu, Sunil J. B., and Huw Williams, "Ukraine Conflict: Ukraine's Electronic Warfare Systems in Focus," Janes, March 4, 2022.

Baraniuk, Chris, "'More Eyes in the Sky.' What War Looks Like in the Era of Accessible Satellites" ["'Більше очей у небі.' Як виглядає війна в епоху доступних супутників"], BBC News Ukraine, March 22, 2022.

Barnes, Julian E., "U.S. Lacks a Clear Picture of Ukraine's War Strategy, Officials Say," *New York Times*, June 8, 2022.

Barnes, Julian E., and Helene Cooper, "Ukrainian Officials Drew on U.S. Intelligence to Plan Counteroffensive," *New York Times*, September 10, 2022.

Barnes, Julian E., Helene Cooper, and Eric Schmitt, "U.S. Intelligence Is Helping Ukraine Kill Russian Generals, Officials Say," *New York Times*, May 4, 2022.

Basmat, Dmytro, "Defense Ministry Working to Limit Satellite Imaging of Ukrainian Territory," *Kyiv Independent*, May 2, 2024.

Baykar, "Fundraising Campaigns for Purchase of Türkiye's Bayraktar TB2 Drone Spreading in West," July 23, 2022.

Bernat, Paweł, "The Evolution of Russian ISR Satellite Capabilities Following the Full-Scale Aggression on Ukraine (2022–2024)," paper presented at the Combating Threats from the Air: Current Issues and Ways to Solve Them conference, Kyiv, Ukraine, October 2024.

Bickerton, James, "U.S. Giving Ukraine Smart Bombs to Double Strike Range Against Russia," *Newsweek*, February 3, 2023.

Biden, Joseph, and Volodymyr Zelenskyy, transcription of remarks given at joint press conference in Fasano, Italy, White House, June 13, 2024.

Bingen, Kari A., Kaitlyn Johnson, and Makena Young, *Space Threat Assessment 2023*, Center for Strategic and International Studies, April 2023.

Bodner, Matthew, "The Long Road to Vostochny: Inside Russia's Newest Launch Facility," *Space News*, January 30, 2019.

Boichak, Olga, and Tetyana Lokot, "Billionaires Won't Save Ukraine's Internet," *Foreign Policy*, November 20, 2022.

Brown, David, "Ukraine Conflict: Where Are Russia's Troops?" BBC News, February 23, 2022.

Bruce, Mary, and Michelle Stoddart, "Biden Gave Ukraine Permission to Strike Inside Russia with US Weapons: Official," ABC News, May 30, 2024.

Bunyak, Valeriia, "GUR: ICEYE Satellite, Purchased with Donations from Ukrainians, Helped Destroy Thousands of Enemy Units" ["ГУР: Супутник ICEYE, придбаний за пожертви українців, допоміг знищити тисячі одиниць ворожої техніки"], Detector Media, March 9, 2023.

Burbach, David T., "Early Lessons from the Russia-Ukraine War as a Space Conflict," Atlantic Council, August 30, 2022.

Burlaka, Oleksandr, "What Russian Satellites See," Universe Space Tech, November 9, 2023.

Cabinet of Ministers of Ukraine, *Public Report of the Head of the State Space Agency of Ukraine for 2019* [*Публічний звіт Голови Державного космічного агентства України за 2019 рік*], February 24, 2020.

Cancian, Mark F., and Chris H. Park, "What Is in the Ukraine Aid Package, and What Does It Mean for the Future of the War?" Center for Strategic and International Studies, May 1, 2024.

Capaccio, Anthony, "Pentagon Deal with Musk's Starlink in Ukraine Extended Six Months for $14 Million," Bloomberg, June 13, 2024.

Chatterjee, Mohar, "Ukraine Scrambles to Keep Internet Up amid Blackouts," Politico, October 21, 2022.

Clark, Joseph, "Space Force General Outlines U.S. Approach to Maintaining Space Superiority," U.S. Department of Defense, March 28, 2024.

Coblentz, Benton, "US Expected to Decide Soon on Long-Range Missiles for Ukraine," Atlantic Council, September 12, 2023.

Connell, Michael, *The Role of Space in Russia's Operations in Ukraine*, CNA, November 2023.

Cooper, Helene, "Ukraine Fires British Long-Range Missiles into Russia," *New York Times*, November 20, 2024.

Copp, Tara, "Elon Musk's Refusal to Have Starlink Support Ukraine Attack in Crimea Raises Questions for Pentagon," Associated Press, September 11, 2023.

Copp, Tara, and Lolita C. Baldor, "Officials: Musk Seeks US Funds for Ukraine Satellite Network," Associated Press, October 15, 2022.

Cranny-Evans, Sam, "Have Russian Cyberattacks Changed the Course of the War in Ukraine?" *National Interest*, June 25, 2022.

Curtis, John, and Claire Mills, *Military Assistance to Ukraine Since the Russian Invasion*, House of Commons Library, March 23, 2022.

"The Defeat of the 'Lancet' of the EW Station 'Bukovel-AD' of the Armed Forces of Ukraine Got into the Frame," Military Review, June 27, 2023.

Defense Intelligence Agency, *Challenges to Security in Space: Space Reliance in an Era of Competition and Expansion*, 2022.

Di Mizio, Giorgio, and Douglas Barrie, "Russian Glide Bombs Add Pressure on Ukraine's Diminishing Air Defences," International Institute for Strategic Studies, March 25, 2024.

Dilanian, Ken, "Biden Administration Walks Fine Line on Intelligence-Sharing with Ukraine," NBC News, March 4, 2022.

Dobrynin, Serhii, and Yevhen Lehalov, "'Relative' of HIMARS: What Can the Ukrainian Missile Complex 'Vilkha' Do" ["'Родич' HIMARS: на що здатний український ракетний комплекс 'Вільха'"], Radio Liberty, January 15, 2024.

DoD—*See* U.S. Department of Defense.

Dovidka.Info, "Communication and Information Retrieval," webpage, undated. As of May 3, 2024: https://dovidka.info/en/communication-and-information-retrieval/

"Dragonfly Aerospace Provided Services to Ukraine in Remote Sensing of the Earth" ["Dragonfly Aerospace надала послуги Україні у дистанційному зондуванні Землі"], Militarnyi, September 19, 2021.

"Dual-Purpose Radar Satellite 'Kondor-E' Deployed to Target Orbit" ["Радиолокационный спутник двойного назначения 'Кондор-Э' выведен на целевую орбиту"], *Vedomosti*, December 19, 2014.

Duffy, Kate, "SpaceX Starlink Has 150,000 Daily Users in Ukraine 5 Weeks After Being Activated, Government Official Says," Business Insider, May 3, 2022.

"Elon Musk Does Not Consider Himself a Traitor, 'Because Congress Did Not Declare War on Russia'" ["Ілон Маск не вважає себе зрадником, 'бо Конгрес не оголошував війну Росії'"], Voice of America, September 11, 2023.

Erwin, Sandra, "As Russia Prepared to Invade, U.S. Opened Commercial Imagery Pipeline to Ukraine," *SpaceNews*, April 6, 2022.

Erwin, Sandra, "Limits on Ukraine's Use of Starlink for War Operations Is a Lesson for U.S. Military," *SpaceNews*, March 9, 2023.

European Space Agency, "Southern Ukraine," webpage, May 1, 2020. As of September 13, 2023: https://www.esa.int/ESA_Multimedia/Images/2020/05/Southern_Ukraine

Evans, Alexandra T., Andrew Radin, Katie Feistel, Krista Langeland, Bruce McClintock, and Howard Wang, *Space Strategic Stability: Assessing U.S. Concepts and Approaches*, RAND Corporation, RR-A2313-1, 2024. As of February 26, 2025:
https://www.rand.org/pubs/research_reports/RRA2313-1.html

"A Failed Lunar Mission Dents Russian Pride and Reflects Deeper Problems with Moscow's Space Industry," Associated Press, August 22, 2023.

Farrow, Ronan, "Elon Musk's Shadow Rule," *New Yorker*, August 21, 2023.

Fedorov, Mykhailo [@FedorovMykhailo], "@elonmusk, while you try to colonize Mars — Russia try to occupy Ukraine! While your rockets successfully land from space — Russian rockets attack Ukrainian civil people! We ask you to provide Ukraine with Starlink stations and to address sane Russians to stand." post on the X platform, February 26, 2022a. As of February 18, 2025:
https://x.com/FedorovMykhailo/status/1497543633293266944?lang=en

Fedorov, Mykhailo, "Always Connected: How Starlink by Elon Musk Is Helping Ukraine During the War" ["Завжди на зв'язку. Чим Україні під час війни допомагає Starlink від Ілона Маска"], RBC-Ukraine, June 17, 2022b.

Feldstein, Steven, "Disentangling the Digital Battlefield: How the Internet Has Changed War," *War on the Rocks*, December 7, 2022.

Finn, Peter, "Ukraine Trying to Take Off into Global Space Industry," *Washington Post*, November 20, 1998.

Foust, Jeff, "SpaceX Worked for Weeks to Begin Starlink Service in Ukraine," *SpaceNews*, March 8, 2022.

Freund, Alexander, "Ukraine Using Starlink for Drone Strikes," Deutsche Welle, March 27, 2022.

Gall, Carlotta, and Vladyslav Golovin, "Some U.S. Weapons Stymied by Russian Jamming in Ukraine," *New York Times*, May 25, 2024.

Gazprom Space Systems, "Infrastructure," webpage, undated. As of February 18, 2024:
https://www.gazprom-spacesystems.ru/en/infrastructure/

Glukhovsky, Mykhailo, "Lieutenant Colonel Honchar: Data from 'Pritula's Satellite' Monopolized by One Entity" ["Підполковник Гончар: Дані з 'супутника Притули' монополізувала одна структура"], *Glavkom*, September 13, 2022.

Gohd, Chelsea, "SpaceX Starlink Satellite Internet Terminals Arrive in Ukraine," Space.com, February 28, 2022.

Gordon, Michael R., Warren P. Strobel, and Vivian Salama, "Biden Administration to Provide Ukraine with More Intelligence, Heavier Weapons to Fight Russia," *Wall Street Journal*, April 13, 2022.

Government Portal of Ukraine, "Ukraine's Participation in the EU Copernicus Program Will Give Impetus to the Development of Key Sectors of Our State, Assures Ivanna Klympush-Tsintsadze," July 10, 2018.

Goward, Dana A., "US Leaders Have Been Warned to Focus on GPS and PNT to Protect the Nation," *SpaceNews*, August 28, 2024.

GPS.gov, "Other Global Navigation Satellite Systems (GNSS)," webpage, October 19, 2021. As of February 18, 2025:
https://www.gps.gov/systems/gnss/

Greig, Jonathan, "CISA's Goldstein: Ukrainian Response to Viasat Hack Proves Need for Redundancy, Resilience," The Record, November 15, 2023.

Hambling, David, "Ukraine Is Spoofing Russian Drones Out of the Sky," *Forbes*, April 21, 2023.

Hambling, David, "Ukraine Fields Unjammable Fiber Optic FPV Attack Drone," *Forbes*, November 7, 2024.

Harris, Shane, Karen DeYoung, Isabelle Khurshudyan, Ashley Parker, and Liz Sly, "Road to War: U.S. Struggled to Convince Allies, and Zelensky, of Risk of Invasion," *Washington Post*, August 16, 2022.

Horton, Alex, and Serhii Korolchuk, "Whatever the Fuss over Elon Musk, Starlink Is Utterly Essential in Ukraine," *Washington Post*, September 18, 2023.

"How Elon Musk's Satellites Have Saved Ukraine and Changed Warfare," *The Economist*, January 5, 2023.

"How Many Objects of the Occupiers Did the 'People's Satellite' ICEYE Help Discover" ["Скільки об'єктів окупантів допоміг виявити 'народний супутник' ICEYE"], Texty.org.ua, March 9, 2023.

Howell, Elizabeth, "Sputnik: The Space Race's Opening Shot," Space.com, September 29, 2020.

Hubashov, Denys, Yevheniia Drozdova, and Nadiia Kelm, "Map of Ukraine's Destruction" ["Карта руйнувань"], Texty.org.ua, March 21, 2023.

Hudkova, Svitlana, "One Cannot Do Without Elon Musk: With Whom and How Ukraine Will Launch Satellites" ["Без Ілона Маска не обійтися: з ким і як Україна буде запускати супутники"], Apostrophe [Апостроф], May 31, 2021.

Hurska, Alla, "Revival of Ukrainian Space Sector: A Viable New Prospect?" *Eurasia Daily Monitor*, Vol. 17, No. 49, April 13, 2020.

Husyev, Hlib, "Following the USSR Trajectory — Ukraine: A Great History of Ukrainian Space, from Independence to the Present Day. People, Factories, and Launch Vehicles" ["За траєкторією СРСР — Україна: велика історія українського космосу, від Незалежності до наших днів. Люди, заводи та ракети-носії"], Babel, September 9, 2019.

ICEYE, "ICEYE Signs Contract to Provide Government of Ukraine with Access to Its SAR Satellite Constellation," press release, August 18, 2022.

"ICEYE and the Ministry of Defence of Ukraine Sign a Memorandum of Cooperation," Defence Industry Europe, July 8, 2024.

Insinna, Valerie, "SpaceX Beating Russian Jamming Attack Was 'Eyewatering': DoD Official," Breaking Defense, April 20, 2022.

Ismay, John, "Russian Guided Weapons Miss the Mark, U.S. Defense Officials Say," *New York Times*, May 9, 2022a.

Ismay, John, "What Are JDAMs? And What Will They Do for Ukraine?" *New York Times*, December 23, 2022b.

Joint Publication 3-14, *Joint Space Operations*, Joint Chiefs of Staff, August 23, 2023.

Jones, Seth G., Riley McCabe, and Alexander Palmer, *Ukrainian Innovation in a War of Attrition*, Center for Strategic and International Studies, February 2023.

Jordan, Dearbail, "Elon Musk Says He Withheld Starlink over Crimea to Avoid Escalation," BBC News, September 8, 2023.

Katsylo, Denys, "We Go Wherever We Can. The Head of the Board of 'Ukrzaliznytsia' Talks About Work During the War and Elon Musk's Starlink" ["Їдемо всюди, де тільки можемо. Голова правління 'Укрзалізниці' – про роботу під час війни та Starlink від Ілона Маска"], *Forbes*, March 2, 2022.

Khurshudyan, Isabelle, Karen DeYoung, Alex Horton, and Karoun Demirjian, "Ukraine Wants More 'Game-Changer' HIMARS. The U.S. Says It's Complicated." *Washington Post*, July 24, 2022.

Khurshudyan, Isabelle, and Alex Horton, "Russian Jamming Leaves Some High-Tech U.S. Weapons Ineffective in Ukraine," *Washington Post*, May 24, 2024.

Khurshudyan, Isabelle, Dan Lamothe, Shane Harrie, and Paul Sonne, "Ukraine's Rocket Campaign Reliant on U.S. Precision Targeting, Officials Say," *Washington Post*, February 9, 2023.

Khurshudyan, Isabelle, Kostiantyn Khudov, Dan Lamothe, and Ellen Francis, "Musk Threatens to Stop Funding Starlink Internet Ukraine Relies on in War," *Washington Post*, October 14, 2022.

Kim, Victoria, Richard Pérez-Peña, and Andrew E. Kramer, "Elon Musk Refused to Enable Ukraine Drone Attack on Russian Fleet," *New York Times*, September 8, 2023.

Korshak, Stefan, "New Gliding Bombs and Precision-Guided Missiles Allow Ukraine to Reach Out and Hit Harder," *Kyiv Post*, March 7, 2023.

Kozlova, Natalia, "Suppliers Are to Blame for the Loss of the Military Satellite 'Persona'" ["В гибели военного спутника 'Персона' виноваты поставщики"], *Russian Gazette* [*Российская газета*], February 11, 2009.

Krasnomovets, Pavlo, "Ukrainian Space Enterprises Have Created Thousands of Rockets and Hundreds of Satellites. Why They Did Not Adapt to a World Without the Cold War" ["Космічні підприємства України створили тисячі ракет та сотні супутників для СРСР. Чому їм виявилося так важко пристосуватися до світу без холодної війни?"], *Forbes*, December 27, 2021.

Kube, Courtney, "Ukraine Uses Long-Range ATACMS Against Russia for the First Time," NBC News, April 24, 2024.

Kuksa, Vitaliy, "Two Faces of Military Communication: How the American Army Adapts Technology to the Soviet Model" ["Два обличчя військового зв'язку. Як в Армії американську техніку підганяють під радянську модель"], Texty.org.ua, May 5, 2016.

Laffaiteur, Michel, Oleg Fedorov, and Sergey Gerasymchuk, *Space Activities: Ukraine Looking for New Developments*, European Space Policy Institute, August 2011.

Lavrut, Oleksandr, Tetiana Valeriivna Lavrut, Oleg Klimovich, and Yuriy Mykolaiovych Zdorenko, "New Technologies and Communication Tools in the Armed Forces of Ukraine: Transformation Path and Development Perspectives" ["Новітні технології та засоби зв'язку у ЗСУ: шлях трансформації та перспективи розвитку"], Ukrainian Military Pages, April 17, 2019.

Lee, Michelle Ye Hee, and Joyce Sohyun Lee, "North Korea May Be Sending Arms to Russia for Ukraine War, Images Suggest," *Washington Post*, October 16, 2023.

Lendon, Brad, "Russia Used an Advanced Hypersonic Missile for the First Time in Recent Strike, Ukraine Claims," CNN, February 13, 2024.

"Lifecell Partially Restored Mobile Communication in Bucha Using Starlink" ["Lifecell частково відновив у Бучі мобільний зв'язок за допомогою Starlink"], Ukrinform, April 8, 2022.

Lima-Strong, Cristiano, "U.S. Quietly Paying Millions to Send Starlink Terminals to Ukraine, Contrary to SpaceX Claims," *Washington Post*, April 8, 2022.

Liptak, Kevin, "Biden Predicts Russia 'Will Move in' to Ukraine, but Says 'Minor Incursion' May Prompt Discussion over Consequences," CNN, January 19, 2022.

Lockheed Martin, "US Army Awards $4.79 Billion Production Contract for All-Weather GMLRS Rocket," press release, April 27, 2023.

Lopez, C. Todd, "HIMARS, Excalibur Rounds Headed for Ukraine in $625 Million Security Assistance Package," U.S. Department of Defense, October 4, 2022.

"Lutsenko on TSN. Tyzhden Showed Evidence of Sabotage at the Warehouses in Balakliia" ["Луценко в ефірі ТСН.Тижня показав докази диверсії на складах у Балаклії"], Television News Service, October 2, 2017.

Luzin, Pavel, "Russia's Space Satellite Problems and the War in Ukraine," *Eurasia Daily Monitor*, Vol. 19, No. 76, May 24, 2022.

Mack, Eric, "US Military Says SpaceX Handily Fought Off Russian Starlink Jamming Attempts," CNET, April 22, 2022.

Maksymenko, Olena, "General Director of Luch Design Bureau Oleh Korostelov: 'The Military Requests 2,000 Missiles, but We Can Only Provide 600 or 800'" ["Гендиректор КБ 'Луч' Олег Коростельов: 'Військові пишуть, що їм потрібно 2000 ракет, а ми їм забезпечуємо тільки 600 або 800'"], Texty.org.ua, September 6, 2021.

Malyasov, Dylan, "Satellite Imagery Shows Hundreds of Russian Tanks near the Border with Ukraine," Defence Blog, August 30, 2020.

Manko, Vitaliy, "Digital Army of Ukraine: Forewarned Means Armed" ["Цифрова армія України: попереджений – значить озброєний"], Censor.net, December 24, 2019.

Marquardt, Alex, "Exclusive: Musk's SpaceX Says It Can no Longer Pay for Critical Satellite Services in Ukraine, Asks Pentagon to Pick Up the Tab," CNN, October 14, 2022.

Marquardt, Alex, Natasha Bertrand, and Zachary Cohen, "Russia's Jamming of US-Provided Rocket Systems Complicates Ukraine's War Effort," CNN, May 6, 2023.

Marquardt, Alex, and Kristin Fisher, "SpaceX Admits Blocking Ukrainian Troops from Using Satellite Technology," CNN, February 9, 2023.

Marrow, Michael, "Pentagon Working with Ukraine, SpaceX to Prevent Russian Exploitation of Starlink," Breaking Defense, April 10, 2024.

Marson, James, and Thomas Grove, "Russia Using Thousands of Musk's Starlink Systems in War, Ukrainian General Says," *Wall Street Journal*, February 15, 2024.

McClintock, Bruce, "The Russian Space Sector: Adaptation, Retrenchment, and Stagnation," *Space and Defense*, Vol. 10, No. 1, Spring 2017.

McClintock, Bruce, and Mélusine Lebret, "Russian Space Strategy and Capabilities: A Tale of Decline," in Alessandro Gili, ed., *The Sky Is Not the Limit: Geopolitics and Economics of the New Space Race*, Ledizioni LediPublishing, 2024.

Melnyk, Tayisa, "Stinging 'Kropyva'. How Ukrainian Software for Artillerymen Affects the Course of the War" ["Жалюча 'Кропива'. Як українське програмне забезпечення для артилеристів впливає на перебіг війни"], *Forbes*, July 24, 2022.

Miller, Christopher, Mark Scott, and Bryan Bender, "UkraineX: How Elon Musk's Space Satellites Changed the War on the Ground," Politico, June 8, 2022.

"The Minister of Defense Urged Ukrainians to Destroy Russian Electronic Warfare and Reconnaissance Systems" ["Міністр оборони закликав українців знищувати російські системи радіоелектронної боротьби і розвідки"], Radio Liberty, March 9, 2022.

Mitzer, Stijn, and Joost Oliemans, "Novel Capabilities: Ukraine's Vilkha Guided MRL," Oryx, May 10, 2021.

Molfar, "Destruction of Electronic Warfare Equipment as a Prelude to a Counteroffensive. Analytics Confirms Dates of Counteroffensive Announced by the Ministry of Defense of Ukraine," undated.

Mozur, Paul, and Aaron Krolik, "The Invisible War in Ukraine Being Fought over Radio Waves," *New York Times*, November 19, 2023.

Murauskaite, Egle E., *U.S. Assistance to Ukraine in the Information Space: Intelligence, Cyber, and Signaling*, Asymmetric Threats Analysis Center, University of Maryland, February 2023.

Musk, Elon [@elonmusk], "There was an emergency request from government authorities to activate Starlink all the way to Sevastopol. The obvious intent being to sink most of the Russian fleet at anchor. If I had agreed to their request, then SpaceX would be explicitly complicit in a major act of war and conflict escalation." post on the X platform, September 7, 2023. As of February 19, 2025: https://x.com/elonmusk/status/1699917639043404146?lang=en

"Musk Says He Refused Kyiv Request for Starlink Use in Attack on Russia," Reuters, September 8, 2023.

Nabozhnyak, Oleksiy, "'Russians Have Hundreds of New Satellite Dishes in Kharkiv Oblast' – Radio and Military Technology Consultant," Slidstvo.Info, May 15, 2024.

National Space Management and Testing Center, "Structure of NSFCTC," webpage, undated. As of September 13, 2023: https://spacecenter.gov.ua/struktura?lang=en

National Reconnaissance Office, "Tri-Seal Commercial Space Protection Framework," August 31, 2023.

Nekrasov, Vsevolod, "A Ukrainian Astronaut on the ISS Is Real. What Will Ukraine's Accession to the European Space Agency Do" ["Український астронавт на МКС – це реально. Що дасть Україні вступ до Європейського космічного агентства"], Ekonomichna Pravda, August 1, 2022.

Observing Systems Capability Analysis and Review Tool, "Satellite: SICH 2-30," webpage, last updated May 16, 2022. As of February 19, 2025: https://space.oscar.wmo.int/satellites/view/sich_2_30

Olearchyk, Roman, "Military Briefing: Russia Has the Upper Hand in Electronic Warfare with Ukraine," *Financial Times*, January 6, 2024.

"On Gagarin Flight Anniversary, Putin Vows Russia Will Remain Space Power," Reuters, April 12, 2021.

Orlova, Violetta, "The Ministry of Digital Transformation of Ukraine Explained How Elon Musk's Starlink Internet Will Help Ukrainians in the War" ["У Мінцифри розповіли, як інтернет Ілона Маска Starlink допоможе українцям на війні"], Unian, March 7, 2022.

Parker, Charlie, "Specialist Ukrainian Drone Unit Picks Off Invading Russian Forces as They Sleep," *The Times*, March 18, 2022.

Pearson, James, and Raphael Satter, "Internet in Ukraine Disrupted as Russian Troops Advance," Reuters, February 27, 2022.

"Persona Satellite Overview," SpaceFlight101.com, undated.

Podvig, Pavel, "Russian Space Systems and the Risk of Weaponizing Space," in Samuel Bendett, Mathieu Boulègue, Richard Connolly, Margarita Konaev, Pavel Podvig, and Katarzyna Zysk, *Advanced Military Technology in Russia: Capabilities and Implications*, Chatham House, September 2021.

"Pokrova EW System Is a Real Game-Changer in Ukrainian Fight Against Shahed-136 Drones and Cruise Missiles, That Renders GPS Receivers Useless," Defense Express, November 4, 2023.

"Poroshenko: We Do Not Need Lethal Weapons from the West" ["Порошенко: нам не потрібна смертельна зброя від Заходу"], BBC News Ukraine, September 25, 2014.

Porter, Jon, "SpaceX Inks First Space Force Deal for Government-Focused Starshield Satellite Network," The Verge, September 28, 2023.

Porter, Tom, "Russia Is Jamming 'Sophisticated' US Weapons Being Used in Ukraine, Making Them Useless, Report Says," Business Insider, August 4, 2023.

"Prytula Provided the Armed Forces of Ukraine Access to the ICEYE Satellite Imaging Database" ["Притула передав ЗСУ доступ до бази даних супутникових знімків ICEYE"], Detector Media, August 18, 2022.

Rao, Rahul, "Russia Launches Heavy-Lift Angara Rocket on 3rd Test Flight, but Misses Intended Orbit: Reports," Space.com, December 30, 2021.

Reach, Clint, Alyssa Demus, Michelle Grisé, Khrystyna Holynska, Christopher Lynch, Dara Massicot, and David Woodworth, *Russia's Evolution Toward a Unified Strategic Operation: The Influence of Geography and Conventional Capacity*, RAND Corporation, RR-A1233-8, 2023. As of December 6, 2024: https://www.rand.org/pubs/research_reports/RRA1233-8.html

Roscosmos [Роскосмос], "Assembly and Testing of the First Satellite of the Sphere Project Has Begun" ["Начались сборка и испытания первого спутника проекта 'Сфера'"], August 21, 2022.

Roth, Andrew, "UK Sends UN Experts Photographs of North Korean Shipments to Russia," *The Guardian*, January 22, 2024.

Roulette, Joey, "SpaceX Curbed Ukraine's Use of Starlink Internet for Drones – Company President," Reuters, February 9, 2023.

Rubin, Uzi, "Russia's Iranian-Made UAVs: A Technical Profile," Royal United Services Institute for Defence and Security Studies, January 13, 2023.

Rudenko, Anzhelika, and Aleksina Dorohan, "'Why Trenches, Are We Expecting a War?' Defensive Fortifications in Crimea: What Can Be Seen from the Satellite?" ["'Навіщо окопи, чекаємо на війну?' Оборонні укріплення в Криму. Що видно із супутника?"], Crimea.Realities, April 5, 2023.

"Russia Launches First Soyuz Rocket Since Space Station Mission Failure," NBC News, June 5, 2015.

"Russia Orbits First Sfera Constellation Satellite," Aviation Week Intelligence Network, October 24, 2022.

"Russia Uses Zircon Hypersonic Missile in Ukraine for First Time, Researchers Say," Reuters, February 12, 2024.

"Russia Using Thousands of SpaceX Starlink Terminals in Ukraine, WSJ Says," Reuters, February 15, 2024.

"Russia Warns West: We Can Target Your Commercial Satellites," Reuters, October 27, 2022.

Satariano, Adam, Scott Reinhard, Cade Metz, Sheera Frenkel, and Malika Khurana, "Elon Musk's Unmatched Power in the Stars," *New York Times*, July 28, 2023.

"Satellite Image of the Aftermath of the Destruction of the S-400 Air Defense System in Crimea Appears" ["З'явився супутниковий знімок наслідків знищення ЗРК С-400 у Криму"], Militarnyi, August 28, 2023.

Satter, Raphael, "Satellite Outage Caused 'Huge Loss in Communications' at War's Outset – Ukrainian Official," Reuters, March 15, 2022.

Schmitt, Benjamin, "Catching Russia Sanctions Busters – From Orbit," Center for European Policy Analysis, June 6, 2023.

Schwirtz, Michael, Scott Reinhard, and Josh Holder, "How Russia Has Increased Its Military Buildup," *New York Times*, January 27, 2022.

Schwirtz, Michael, Anton Troianovski, Yousur Al-Hlou, Masha Froliak, Adam Entous, and Thomas Gibbons-Neff, "Putin's War," *New York Times*, December 16, 2022.

Segura, Cristian, "Ukraine Places Trust in White House to Prevent Musk Pulling Plug on Starlink Comms for the Armed Forces," *El País*, September 19, 2023.

Shinkman, Paul D., "Ukraine Has Put 'Fear into the Russian Pilots,' Top U.S. General Says," U.S. News & World Report, June 7, 2023.

Sholudko, V. G., M. Yu. Esaulov, O. V. Vakulenko, T. G. Gursky, and M. M. Fomin, *Organization of Military Communication: Training Manual* [*Організація військового зв'язку: навчальник посібник*], Ministry of Defence of Ukraine, 2017.

Sitnikova, Iryna, "The Ukrainian Satellite 'Sich 2-30' Works in a Limited Mode, It Cannot Transmit Images – the Head of the State Space Agency" ["Український супутник 'Січ 2-30' працює в обмеженому режимі, він не може передавати знімки – голова Держкосмосу"], Hromadske, August 23, 2022.

Skove, Sam, "How Elon Musk's Starlink Is Still Helping Ukraine's Defenders," Defense One, March 1, 2023a.

Skove, Sam, "Russia Is Jamming US Precision Weapons in Ukraine, US General Says," Defense One, December 12, 2023b.

Skove, Sam, "Another US Precision-Guided Weapon Falls Prey to Russian Electronic Warfare, US Says," Defense One, April 28, 2024.

Soesanto, Stefan, *The Ukrainian Way of Digital Warfighting: Volunteers, Applications, and Intelligence Sharing Platforms*, Center for Security Studies, July 2024.

Sonne, Paul, Isabelle Khurshudyan, Serhiy Morgunov, and Kostiantyn Khudov, "Battle for Kyiv: Ukrainian Valor, Russian Blunders Combined to Save the Capital," *Washington Post*, August 24, 2022.

SpaceX, "Starshield," webpage, undated. As of February 19, 2024:
https://www.spacex.com/starshield/index.html

"SpaceX Relinquishes Control of 'Ukrainian' Starlink Terminals to Pentagon," Militarnyi, September 14, 2023.

"Special Topic. Military Communication: Transformation" ["Спецтема. Військовий зв'язок: трансформація"], Defense Express, June 7, 2018.

Srivastava, Mehul, Roman Olearchyk, Felicia Schwartz, and Christopher Miller, "Ukrainian Forces Report Starlink Outages During Push Against Russia," *Financial Times*, October 7, 2022.

Starlink, "High-Speed Internet Around the World," webpage, undated. As of April 24, 2024:
https://www.starlink.com/us

"Starlink: Why Elon Musk's Internet Is so Important for Ukraine" ["Starlink: чому інтернет від Ілона Маска такий важливий для України"], BBC News Ukraine, October 19, 2022.

"Starlink's Performance in Ukraine Has Ignited a New Space Race," *The Economist*, January 5, 2023.

State Kyiv Design Bureau "Luch," homepage, undated. As of May 8, 2024:
https://www.luch.kiev.ua/en/

State Space Agency of Ukraine, "In Ukraine, the Regional Center of the European Union 'Copernicus' Program Has Started Operating" ["В Україні розпочав роботу Регіональний центр програми Європейського Союзу 'Copernicus'"], January 28, 2020.

Stepanenko, Kateryna, Karolina Hird, Riley Bailey, Angelica Evans, and Frederick W. Kagan, *Russian Offensive Campaign Assessment, January 13, 2024*, Institute for the Study of War, January 13, 2024.

Stoll, Hunter, John Hoehn, and William Courtney, "Air Defense Shapes Warfighting in Ukraine," RealClearDefense, February 22, 2024.

Stone, Mike, "Ukraine's New 100-Mile Bomb from Boeing Is Ready, Sources Say," Reuters, January 30, 2024.

Stone, Mike, and Joey Roulette, "SpaceX's Starlink Wins Pentagon Contract for Satellite Services to Ukraine," Reuters, June 1, 2023.

Strobel, Warren P., and Robert Wall, "Ukraine War Puts Spy Satellites for Hire in the Spotlight," *Wall Street Journal*, May 1, 2022.

Sukhankin, Sergey, "Blind, Confuse and Demoralize: Russian Electronic Warfare Operations in Donbas," Jamestown Foundation, August 27, 2021.

Swope, Clayton, Kari A. Bingen, Makena Young, Madeleine Chang, Stephanie Songer, and Jeremy Tammelleo, *Space Threat Assessment 2024*, Center for Strategic and International Studies, April 2024.

"A Technophile's Guide to the Evolution of Russian Shahed Drones," *Kyiv Post*, December 10, 2023.

Tereshchuk, Halyna, "Space Wars. Is Russia Capable of Using Nuclear Weapons in Space" ["Космічні війни. Чи здатна Росія застосувати ядерну зброю у космосі"], Radio Liberty, March 2, 2024.

Thomas, Timothy, *Russia's Electronic Warfare Force: Blending Concepts with Capabilities*, MITRE Center for Technology and National Security, September 2020.

Thornton, Rod, and Marina Miron, *Russian Military Space-Related Capabilities: The Vital Deterrence Role of Counterspace Weapons (Part I)*, Freeman Air and Space Institute, May 2024.

Tingley, Brett, "Russia Says Private Satellites Could Become 'Legitimate Target' During Wartime," Space.com, September 16, 2022a.

Tingley, Brett, "White House Says US Would Respond If Russia Targets Commercial Satellites," Space.com, October 28, 2022b.

Tokar, Ihor, "Declassified. How Does Ukraine Monitor Russia in Crimea from Space?" ["Розсекретили. Як Україна стежить за Росією в Криму з космосу?"], Radio Liberty, February 7, 2021.

Tong, Sebastian, "Musk Says SpaceX Has Spent $80 Million Out of Pocket to Support Ukraine," Bloomberg, October 3, 2022.

Torrieri, Marisa, "How Satellite Imagery Magnified Ukraine to the World," *Via Satellite*, October 24, 2022.

Triebert, Christiaan, Blacki Migliozzi, Alexander Cardia, Muyi Xiao, and David Botti, "Fake Signals and American Insurance: How a Dark Fleet Moves Russian Oil," *New York Times*, May 30, 2023.

Trofimov, Yaroslav, "High-Tech American Weapons Work Against Russia—Until They Don't," *Wall Street Journal*, July 10, 2024.

Trofimov, Yaroslav, Micah Maidenberg, and Drew FitzGerald, "Ukraine Leans on Elon Musk's Starlink in Fight Against Russia," *Wall Street Journal*, July 16, 2022.

"Two Years of the Russian-Ukrainian War Through the Lens of Satellites," Universe Space Tech, February 23, 2024.

"Ukraine Launched a Massive Strike Using Rare Vilkha Rockets, Russia Claims," Defense Express, January 3, 2024.

"Ukraine's Military Space. Part 2" ["Військовий космос України. Частина 2"], Militarnyi, December 15, 2017.

"Ukrainian Bukovel-AD EW System 'Landed' Russian ZALA 421-16E2 UAV," Militarnyi, March 16, 2023.

"Ukrainian Bukovel-AD R4 EW-Systems Deployed near the Border," InformNapalm, November 21, 2021.

"Ukrainian Sich-2-30 Satellite Successfully Launched into Orbit," Defense Express, January 14, 2022.

"Ukrainian Space Center Showed How Russia Is Increasing Its Military Presence in Crimea" [В українському космічному центрі показали, як Росія нарощує військову присутність у Криму"], Radio Liberty, February 8, 2021.

"Ukrainian Tactical Missiles. Instead of Designing Until 2020, It's Better to Buy Them from the US Right Now" ["Українські тактичні ракети. Замість проектувати до 2020 року - краще їх купити у США прямо зараз"], Texty.org.ua, January 22, 2010.

United Kingdom Ministry of Defence 🇬🇧 [@DefenceHQ], "Latest Defence Intelligence update on the situation in Ukraine – 6 December 2023. Find out more about Defence Intelligence's use of language: ow.ly/Il6C50QfRHV 🇺🇦 #StandWithUkraine 🇺🇦," post on the X platform, December 6, 2023. As of February 19, 2025: https://x.com/DefenceHQ/status/1732340880638406852

United Nations Development Programme, "63% of Ukrainians Use State E-Services, User Numbers Grow for Third Year in Row – Survey," press release, January 25, 2023.

U.S. Agency for International Development, "USAID Safeguards Internet Access in Ukraine Through Public-Private-Partnership with SpaceX," press release, April 5, 2022.

U.S. Department of Defense, "Fact Sheet on U.S. Security Assistance for Ukraine (Roll Up) as of April 21, 2022," press release, April 22, 2022a.

U.S. Department of Defense, "$600 Million in Additional Security Assistance for Ukraine," press release, September 15, 2022b.

U.S. Department of Defense, *Commercial Space Integration Strategy*, 2024.

U.S. Department of Defense, "Biden Administration Announces Additional Security Assistance for Ukraine," press release, May 9, 2023.

U.S. Government Accountability Office, *GPS Alternatives: DoD Is Developing Navigation Systems but Is Not Measuring Overall Progress*, GAO-22-106010, August 2022.

U.S. Government Accountability Office, *GPS Modernization: Space Force Should Reassess Requirements for Satellites and Handheld Devices*, GAO-23-106018, June 2023.

U.S. Space Force, *United States Space Force Vision for Satellite Communications (SATCOM)*, January 23, 2020.

U.S. Space Force, *U.S. Space Force Commercial Space Strategy: Accelerating the Purposeful Pursuit of Hybrid Space Architectures*, April 8, 2024.

USSF—*See* U.S. Space Force.

van Brugen, Isabel, "Russia Threatens to Shoot Down Western Satellites," *Newsweek*, October 19, 2023.

Viasat, "KA-SAT Network Cyber Attack Overview," March 30, 2022.

Vigliarolo, Brandon, "Russia Says Starlink Satellites Could Become Military Targets," *The Register*, October 28, 2022.

"Vilkha-M Hit a Target at 110 Km with a Deviation of 15 M" ["Вільха-М вразила ціль на 110 км з відхиленням у 15 м"], Ukrainian Military Pages, April 26, 2020.

Vinion, Kurt, "How Elon Musk's Starlink Became Invaluable to Ukraine's War Effort," Radio Free Europe/Radio Liberty, October 20, 2022.

Wall, Mike, "Russia Launches Final GLONASS-M Navigation Satellite into Orbit," Space.com, November 30, 2022.

Wall, Mike, and Meghan Bartels, "Laika the Space Dog: First Living Creature in Orbit," Space.com, January 14, 2022.

"The War in Ukraine Shows How Technology Is Changing the Battlefield," *The Economist*, July 3, 2023.

Wang, Howard, Gregory Graff, and Alexis Dale-Huang, *China's Growing Risk Tolerance in Space: People's Liberation Army Perspectives and Escalation Dynamics*, RAND Corporation, RR-A2313-2, 2024. As of February 19, 2025: https://www.rand.org/pubs/research_reports/RRA2313-2.html

Wattles, Jackie, "SpaceX Sent Starlink Internet Terminals to Ukraine. They Could Paint a 'Giant Target' on Users' Backs, Experts Say," CNN, March 4, 2022.

Wendling, Mike, "US Lawmakers Quiz Musk's Starlink over Russia Claims," BBC News, March 7, 2024.

"What Do We Know About the Ukrainian New Year's Eve Attack on Russian Troops?" Reuters, January 3, 2023.

"What High-Precision Weapons Can Ukraine Receive from NATO" ["Яку високоточну зброю Україна може отримати від НАТО"], Espreso.TV, September 5, 2014.

"What Is the ICEYE Satellite Purchased with Bayraktar TB2 Funds and Why It Really Makes a Difference," Defense Express, August 19, 2022.

"What Pentagon Leaders Say They Have Learned One Year on from the Battle in Ukraine," Breaking Defense, February 23, 2023.

"When GPS Fails, How Can Weapons Find Their Targets?" *The Economist*, July 14, 2023.

Wilson, Joseph, "Musk's Plan to End Russian War Infuriates Ukraine on Twitter," Associated Press, October 4, 2022.

Withington, Thomas, "Jamming JDAM: The Threat to US Munitions from Russian Electronic Warfare," Royal United Services Institute for Defence and Security Studies, June 6, 2023.

Wood, Graeme, "A Suspicious Pattern Alarming the Ukrainian Military," *The Atlantic*, March 18, 2024.

Yarmolenko, Oleksiy, "Germany Will Allocate €20 Million to Pay for 'Starlink' Services in Ukraine," Babel, January 10, 2023.

Yurasov, Stas, "Starlink by Elon Musk Postpones Launch Date of Satellite Internet Service in Ukraine to 2023" ["Starlink Ілона Маска переніс дату запуску послуги супутникового інтернету в Україні на 2023 рік"], dev.ua, February 21, 2022.

Zabrodskyi, Mykhaylo, Jack Watling, Oleksandr V. Danylyuk, and Nick Reynolds, *Preliminary Lessons in Conventional Warfighting from Russia's Invasion of Ukraine: February–July 2022*, Royal United Services Institute for Defence and Security Studies, November 30, 2022.

Zak, Anatoly, "Kobalt-M Reconnaissance Satellite Series," webpage, RussianSpaceWeb.com, September 11, 2023. As of February 19, 2025: https://www.russianspaceweb.com/kobalt_m.html

Zak, Anatoly, "Soyuz Launches Third Bars-M Satellite," webpage, RussianSpaceWeb.com, January 15, 2024. As of February 19, 2025: https://www.russianspaceweb.com/bars-m3.html

Zaluzhny, Valery, "The Commander-in-Chief of Ukraine's Armed Forces on How to Win the War," *The Economist*, November 1, 2023.

Zegart, Amy, "Open Secrets: Ukraine and the Next Intelligence Revolution," *Foreign Affairs*, Vol. 102, No. 1, January/February 2023.

Zetter, Kim, "Viasat Hack 'Did Not' Have Huge Impact on Ukrainian Military Communications, Official Says," Zero Day, September 26, 2022.

Zhirohov, Mikhail, "Ukraine Grows EW Capability to Address Russia in Donbass," Janes, June 12, 2020.

www.ingramcontent.com/pod-product-compliance
Lightning Source LLC
Chambersburg PA
CBHW080856150625
28215CB00014B/113
* 9 7 8 1 9 7 7 4 1 5 2 1 9 *